Synthesis Lectures on Green Energy and Technology

Series Editor

Zhongchao Tan, Department of Mechanical and Mechatronics Engineering, University of Waterloo, Waterloo, Canada

This series publishes short books in environmental-friendly and sustainable technologies, in particular state-of-the-art research and development in green energy and pollution control. The books in this series are written by top researchers and professionals. The series addresses researchers, advanced students, technical consultants as well as decision makers in industries and politics. Hence, the level of presentation spans from instructional to highly technical.

Fangyan Sun

Synthesis and Device Applications of Graphene Derivatives and Quantum Dots

Springer

Fangyan Sun
Postdoctoral Researcher
University of Windsor
Windsor, Canada

University of Waterloo
Waterloo, ON, Canada

ISSN 2948-2720 ISSN 2948-2739 (electronic)
Synthesis Lectures on Green Energy and Technology
ISBN 978-3-031-56402-4 ISBN 978-3-031-56403-1 (eBook)
https://doi.org/10.1007/978-3-031-56403-1

This Springer imprint is published by the registered company Springer Nature Switzerland AG
The registered company address is: Gewerbestrasse 11, 6330 Cham, Switzerland

Preface

Nanotechnology and materials science are advancing at a rapid pace, and the synthesis and application of new materials have become key to advancing various technological frontiers. Among these materials, graphene and its derivatives, including graphene oxide (GO), reduced graphene oxide (RGO), and graphene quantum dots (GQDs), have garnered significant attention. Their unique electrical, thermal, and mechanical properties make them ideal candidates for device applications, particularly in the field of electronics and energy.

This book is a comprehensive guide for students, researchers, and professionals interested in the synthesis of graphene-based materials and their incorporation into modern devices. It aims to bridge the gap between fundamental research and practical application, offering a detailed exploration of the processes involved in synthesizing these materials and the intricacies of their integration into device architectures.

This book begins with an introduction to the background of graphene and its derivatives, followed by an extensive literature review for understanding current research and developments in the field. The subsequent chapters focus on the synthesis of graphene-based materials in detailed experimental procedures, material characterizations, and result discussions. The book also focuses on the practical device applications of these synthesized materials. Through detailed chapters on the various uses of graphene, GO, RGO, and GQDs in devices such as MOS devices, solar cells, and photodetectors, readers will gain insights into how these materials are revolutionizing technology.

This book is to not only provide informative resource but also inspires innovation and further exploration in the field. The book is structured to those new to the topic while providing enough depth to be of value to experienced researchers. Hope this book will contribute to the growth of knowledge in nanomaterials and stimulate further research and development in graphene-based technologies.

Waterloo, Canada Fangyan Sun

Contents

1 An Introduction: Solar Cell Technology 1
- 1.1 Fundamental Background of Solar Energy 1
- 1.2 Structural Outline of the Book 2

2 Comprehensive Review: Synthesis, Properties and Applications of Graphene-Based Materials 5
- 2.1 Exploration of Graphene's Structural and Electrical Properties 5
- 2.2 Graphene Oxide 6
  - 2.2.1 Chemical Synthesis of Graphene Oxide 8
  - 2.2.2 Reduction of Graphene Oxide 10
  - 2.2.3 Reduction Mechanism of RGO 12
  - 2.2.4 Work Function of RGO 14
- 2.3 Graphene Quantum Dots 15
  - 2.3.1 Methodologies for Synthesizing GQDs 15
  - 2.3.2 Optical Properties of GQDs 17
- 2.4 Potential Applications of Graphene and Its Derivatives in Device Engineering 18
  - 2.4.1 MOS Devices 18
  - 2.4.2 Graphene-Based Materials as Transparent Conductive Electrodes (TCEs) 19
  - 2.4.3 Role of Graphene-Based Materials in Solar Cell Active Layers 20
  - 2.4.4 Light Absorption in Solar Cells via Graphene Quantum Dots 21
  - 2.4.5 Graphene Quantum Dots-Based Photodetectors 21

3 Synthesis of Graphene Oxide for Solar Cell Applications 23
- 3.1 Synthesis and Characterization of Graphene Oxide 23
  - 3.1.1 Materials Characterizations 24
  - 3.1.2 Measurement of Yield 25
- 3.2 Results and Discussion for Materials Characterizations 25
  - 3.2.1 Morphological and Structural Characterizations 25
  - 3.2.2 Crystallinity Analysis 26

3.2.3 Photoluminescence Analysis 27
3.2.4 Chemical Composition Analysis 28
3.3 Summary 29

4 From Graphene Oxide to Reduced Graphene Oxide 31
4.1 Experimental Process of Reduced Graphene Oxide 31
4.1.1 Formation of GO Thin Film 31
4.1.2 Green Reduction of GO: Ascorbic Acid and HI Methods 31
4.1.3 Fabrication of Free-Standing RGO Films 32
4.2 Material Characterization and Discussion 33
4.2.1 Visual and Surface Analysis 35
4.2.2 AFM Analysis of GO and RGO 36
4.2.3 FTIR Analysis of GO and RGO 38
4.2.4 Structural Analysis of GO and RGO 39
4.2.5 Chemical Composition of GO and RGO Thin Film 40
4.2.6 Electrical Conductivity of RGO and Optimization 41
4.2.7 Work Function of RGO 44
4.2.8 Photoluminescence Analysis of RGO 47
4.3 Summary 47

5 Device Application of Reduced Graphene Oxide 49
5.1 RGO in MOS Devices: Design and Functionality 49
5.2 Fabrication Process of MOS Devices with RGO 49
5.2.1 MOSCAP Device Fabrication Procedure 49
5.2.2 Fabrication Process of RGO as Gate Metal 50
5.2.3 Fabrication Process of Al as Gate Metal 50
5.2.4 Characterization Techniques 51
5.3 Device Characterization and Discussion 51
5.3.1 Capacitance–Voltage (CV) Measurement of MOSCAP 51
5.3.2 Structural Analysis of RGO 52
5.3.3 Chemical Compositions of RGO 53
5.3.4 Electrical Conductivity of RGO 54
5.3.5 Work Function of RGO 54
5.4 Summary 56

6 Synthesis of Graphene Quantum Dots for Photovoltaics 57
6.1 Process Development for Top-Down Synthesized GQD 57
6.1.1 Materials Characterization Techniques 58
6.2 Materials Characterization Results and Discussion 59
6.2.1 Photoluminescence of GQDs 59
6.2.2 Chemical Composition of GQDs 60
6.2.3 Long-Term Aging Test of GQDs 60

6.2.4 Graphitic Structure Analysis ... 62
6.2.5 Internal Factors Dominating Photoluminescence ... 64
6.3 pH-Controlled Synthesis of GQDs: Tailoring for Device Compatibility ... 64
6.3.1 GQD Thin Film Fabrication ... 64
6.3.2 Photoluminescence of Neutralized GQD ... 66
6.3.3 Lifetime Analysis ... 66
6.3.4 Chemical Composition of GQD Film ... 68
6.4 Summary ... 68

7 Top-Down Synthesized Graphene Quantum Dots in Photovoltaic Devices ... 69
7.1 Concept of the GQDs-Based Photodetectors ... 69
7.2 Development of Planar GQDs-Based Photodetectors ... 70
7.2.1 Band Structural Analysis for GQD-Based Photodetector ... 71
7.2.2 Fabrication Process of GQD-Based Photodetectors ... 71
7.3 Planar Heterojunction GQD-Based Photodetector ... 72
7.4 Planar Heterojunction PEG-Passivated GQDs Photodetectors ... 73
7.4.1 Fabrication of PEG-Passivated GQDs Photodetectors ... 73
7.4.2 Device Characterization Techniques ... 74
7.5 Results and Discussion of GQDs-Based Photodetectors ... 75
7.5.1 Optical Properties of PEG-Passivated GQDs Film ... 75
7.5.2 Photodetector Devices Morphology ... 76
7.5.3 Carrier Transport Mechanism ... 77
7.5.4 Current-Voltage (I-V) Response to UV Exposure ... 78
7.5.5 Current-Voltage (I-V) Response to Visible Illumination ... 80
7.6 Summary ... 80

8 Conclusions and Outlook ... 83
8.1 Conclusions ... 83
8.2 Outlook and Future Work ... 85

Bibliography ... 87

An Introduction: Solar Cell Technology 1

1.1 Fundamental Background of Solar Energy

The world currently relies on carbon sources to meet its energy needs; main sources include oil, coal, and natural gas. However, these non-renewable sources have adverse effects on the environment and public health [1]. Solar energy is one of the main renewable sources of energy; it is considered the most promising solution to the energy crisis for its abundance and environmental friendliness. Solar energy is harnessed via photovoltaic (PV) devices, which convert clean solar energy to electricity.

In the domain of photovoltaics, researchers must address the challenge of increasing efficiency while decreasing costs. To achieve this goal, scientists and engineers have found diverse solutions through exploring new materials and improving device structures. With technological development, solar cell applications have become widespread in the military, space, business, agriculture, communication, and public facilities. However, further research and development are needed to enable the large-scale commercialization of solar energy.

Crystalline silicon solar cells are the ancestors of all modern photovoltaic devices; their current efficiency is 20% or higher for commercial solar cells [2]. Although silicon solar cells are leading the PV market, their rigidity, fragility, and high costs prevent them from implementation. This led to advances in the second generation thin-film solar cells, including cadmium telluride (CdTe), Copper indium gallium selenide (CIGS), and amorphous silicon (α-Si). Over the past 20 years, these cells have shown substantial improvement in efficiency, reliability, and cost [3]. Owing to its unusual properties and cost-effectiveness, the multi-junction of α-Si has caught growing interest among the research community. Although the second-generation solar cells perform much better than the first-generation mainstream crystalline silicon solar cells, efficiencies must be further improved and the material costs must be removed [4].

F. Sun, *Synthesis and Device Applications of Graphene Derivatives and Quantum Dots*,
Synthesis Lectures on Green Energy and Technology,
https://doi.org/10.1007/978-3-031-56403-1_1

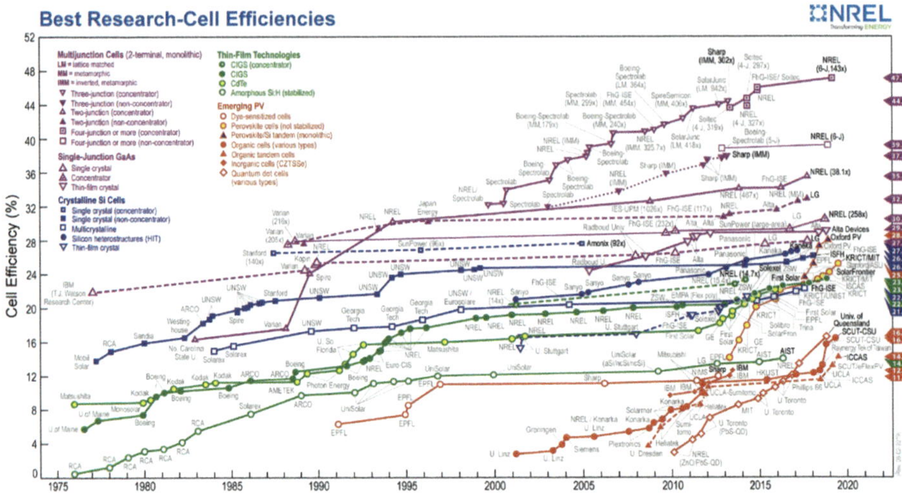

Fig. 1.1 Power conversion efficiencies of worldwide best research solar cells from 1976 to 2020 [5]. This plot is courtesy of the National Renewable Energy Laboratory, Golden, CO

Perovskite solar cells (PSCs) and quantum dot solar cells (QDSCs) represent third-generation solar cells. Perovskites have shown great potential as a light-harvesting and carrier transport material, with perovskite-based cells showing the efficiency of over 20%. However, their drawbacks include poor stability and reliability. Pb-based perovskite materials can achieve high-efficiency performance (PCE > 20%). However, the strong toxicity makes them environmentally unfriendly.

Quantum dots (QDs), with their tunability, high emission, and other attractive properties, are a promising alternative for the future of photovoltaics. Figure 1.1 [5] depicts the power conversion efficiencies for the solar cells reported worldwide. Since their emergence, QDSCs have seen a substantial rise in efficiency to 10% or higher.

1.2 Structural Outline of the Book

This book focuses on the synthesis of graphene-based nanomaterials and their applications in photodetectors and photovoltaic devices. Specifically, scalable synthesis of graphene and its derivatives, device design, fabrication, and testing based on the synthesized nanomaterials are discussed. The overall research structure is shown in Fig. 1.2. The first principal component of this research is the development of novel, scalable, high-throughput chemical synthesis processes for graphene-based materials. These optimized materials, which include graphene oxide (GO), reduced graphene oxide (RGO), and graphene quantum dots (GQDs), are then utilized in the fabrication of photodetectors

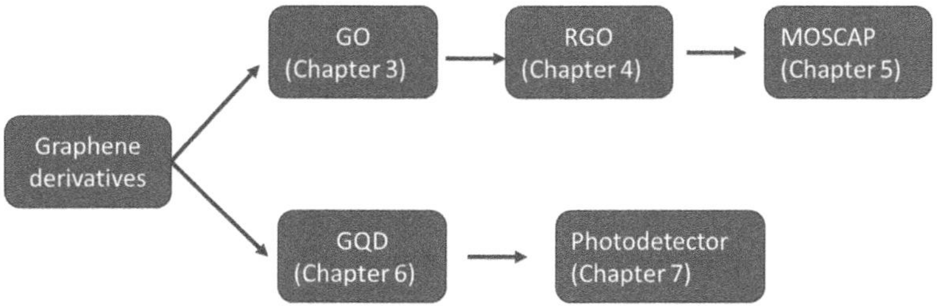

Fig. 1.2 Overall research of graphene-based materials synthesis and application

and photovoltaic devices, highlighting the versatility and potential of graphene derivatives in advanced technological applications. The book also details the specific uses of these materials, such as employing RGO in MOSCAP devices as the gate metal and utilizing GQDs as active layers in photodetectors. This approach highlights the broad spectrum of applications for graphene-based materials, from enhancing the efficiency of photovoltaic devices to advancing the functionality of photodetectors. This work aims to provide readers with a thorough understanding of the synthesis and application of graphene derivatives, paving the way for future innovations in nanotechnology and renewable energy solutions.

Chapter 2 offers a thorough literature review, exploring the structural, physical, and chemical properties, and synthesis methods of graphene, graphene oxide (GO), reduced graphene oxide (RGO), and graphene quantum dots (GQDs). It also delves into the development and current state of MOS devices, graphene-based photovoltaic devices, and photodetectors.

Chapter 3 focuses on the synthesis of graphene oxide (GO), which is prepared as the precursor for most graphene-based materials' chemical synthesis. This chapter primarily concerns the development of GO. This chapter emphasizes the development of GO, covering detailed sections on its synthesis and characterization. It includes comprehensive insights into the results and discussions surrounding GO synthesis, with characterization focusing on microstructural, chemical, and optical analyses.

Chapter 4 discusses the chemical reduction of GO to produce RGO, a process that removes oxygen-containing groups and repairs atomic-scale lattice defects, restoring the graphene lattice's conjugated network structure. It explores a novel two-step reduction process using ascorbic acid (L-AA) and HI, aimed at recovering the electrical conductivity and other intrinsic properties of graphene. It includes a comprehensive overview of RGO synthesis, detailing the synthesis procedures, thin-film fabrication methods, and characterization of RGO's structure and chemical composition, comparing it with GO.

Chapter 5 presents the application of RGO used as the top contact in the MOS devices. Section 5.1 introduces the design of the MOS device structure with an RGO electrode, and Sect. 5.2 gives the fabrication process of MOS devices. Section 5.3 mainly characterizes the MOS devices with capacitance–voltage (C-V) measurement to show the quality of the devices and RGO materials characterizations. Section 5.4 concludes the performance of MOSCAP devices using RGO as gate metal with matched work function and high conductivity. This section highlights how C-V measurements reveal information about charge distribution within the semiconductor, the oxide layer-semiconductor interface, and the semiconductor-metal interface. It explains the influence of surface potential and frequency on charge capacitance distribution, noting distinct responses at high (>1 MHz) and low frequencies to gate voltage changes.

Chapter 6 introduces the modified process for GQD synthesis. GQDs were produced using solution chemistry methods with pH control. In the synthesis route, a facile wet-chemical top-down method is used to produce GQDs from the mixed acidic treatment on graphite. Adjusting the oxidative cutting temperature and pH in an aqueous solution, size tuning was accomplished. The process enables size tuning by adjusting the oxidative cutting temperature and pH in an aqueous solution, highlighting the adaptability of GQD size and properties. This chapter delves into the characterization of the GQDs, focusing on their photoluminescence (PL) properties, and discusses how modifications in pH lead to GQDs with high PL intensity and the fabrication of uniform GQD films. This synthesis approach showcases the potential of GQDs for various applications, leveraging their unique properties through controlled chemical processes.

Chapter 7 focuses on graphene-based photodetectors, exploring the design, fabrication, and characterization of simple and patterned photodetector devices using synthesized GQDs as active layers. This chapter highlight the potential of GQDs as photosensitized layers in photodetectors, particularly for light detection applications.

Chapter 8 provides a comprehensive conclusion and outlook on the research of graphene-based materials, summarizing key findings in both materials' synthesis and applications, and envisioning future directions and potential advancements in the field of graphene-based nanomaterials and their applications in various technologies.

2 Comprehensive Review: Synthesis, Properties and Applications of Graphene-Based Materials

2.1 Exploration of Graphene's Structural and Electrical Properties

Graphene, a two-dimensional carbon nanomaterial with honeycomb structure, represents the simplest form of carbon atoms arranged in sp^2 hybridization. Its single-atom thickness allows graphene to assume various forms: stacking into 3D graphite, rolling into 1D carbon nanotubes, and cutting into 0D graphene quantum dots [6].

This specific structure gives graphene excellent electronic, optical, thermal, and mechanical properties [7]. Graphene is a semimetal and the electronic structure is a zero band gap with a conical structure of valance band and conduction band meeting at the Dirac points [8]. Due to this electronic structure, graphene has a high electronic conductivity up to 35,100 S/m, and carrier mobility at about 2×10^5 $cm^2/(V{\cdot}s)$ at electron densities (n) of 2×10^{11} cm^2 [9, 10]. Moreover, it also has a high thermal conductivity at about 6,000 W mK^{-1} because of the strong stacking of carbon atoms [11]. Due to the one-atom thickness, the transparency of the monolayer graphene is at about 97% [12]. Furthermore, the excellent mechanical properties result from the strong σ-bonds in the graphene sheets. Young's modulus of graphene is about 1 TPa, and fractural strength (about 130 GPa) slightly changes the electrical conductivity of graphene [13].

There are several methods to synthesize graphene materials. The first graphene was prepared from graphite by scotch tape in 2004 [14]. This method can produce high-quality graphene at a low cost but low yield. From then on, graphene synthesis research was stimulated and classified into mechanical exfoliation/cleavage of graphite [14], synthesis of SiC [15], chemical vapor deposition (CVD) [16], oxidation–reduction reaction [17], and so on. Graphene epitaxially grown on SiC by vacuum graphitization is the alternative method of mechanical exfoliation. The number of layers can be controlled for high-quality products. However, the size of graphene is limited by SiC wafers and with high costs

F. Sun, *Synthesis and Device Applications of Graphene Derivatives and Quantum Dots*,
Synthesis Lectures on Green Energy and Technology,
https://doi.org/10.1007/978-3-031-56403-1_2

[15]. CVD method is another choice to produce high-quality graphene, and it is often applied to fabricating electronic devices. For example, the source sucrose or polymethyl methacrylate (PMMA) can grow large single-layered graphene on Cu or Ni substrates at 800–1000 °C [18]. However, the CVD method uses complex apparatus with a high energy cost, and it is time-consuming to grow a continuous film. In addition, high-temperature synthesis and subsequent transfer processes bring defects in the graphene structure.

Alternatively, the oxidation–reduction reaction can synthesize graphene derivates: graphene oxide (GO), reduced graphene oxide (RGO), and so on. This method is a widely used solution-phase method with low cost, high yield, and mass production. The oxidization process produces water-soluble GO, which can be separated into layers by sonication. The reduction process can reduce GO to RGO with recovered graphitic lattice and functional groups elimination to improve the electrical conductivity beneficial for electronic applications. Chinese graphene manufacturers mostly use this method to produce small graphene sheets [6]. However, oxidation often produces toxic gases and low production efficiency. Moreover, the final products are usually thick and not uniform, and these problems need to be solved.

2.2 Graphene Oxide

GO is strongly related to graphene because it is mostly the precursor for graphene. GO can be chemically or thermally converted into reduced graphene oxide (RGO) in the oxidation–reduction process. The reduction of GO is essential to producing high-quality graphene. GO is a derivative of carbon, presenting as a wrinkled one-atom-thick sheet with a thicknesless than 1 nm thickness s of less than 1 nm in two dimensions. It has sp^2 and sp^3 bonds with oxygenated functional groups, including hydroxyl groups (OH), epoxy groups (C–O–C), carbonyl groups (C = O), and carboxyl groups (COOH), on the edges and surface of its graphene skeleton [19].

GO has a similar lattice structure to graphite oxide, but it is single- or multi-layered sheets. However, the precise atomic and electronic structure of GO is still unknown. Nowadays, the Lerf-Klinowski model is widely adopted based on NMR results in Fig. 2.1. This model is defined into two parts: oxygenated functional groups and sp^2- and sp^3-hybridized carbon atoms [20]. The hydroxy and epoxy functional groups are attached at the basal plane, separated by carboxylic acids, which are lightly oxidized on the edges [20]. Depending on the synthesis method, the structure of GO varies with the pH of the oxidation conditions, graphite precursors, experimental setup, and so on. Due to these functional groups, GO presents acidic properties.

There are multiple technologies for GO characterization. The thickness of GO, which is approximately 1 nm, can be accurately measured using techniques such as Nuclear Magnetic Resonance (NMR) or Atomic Force Microscopy (AFM), to calculate the number of layers [22]. Scanning tunnelling microscopy (STM) can compare pristine graphene

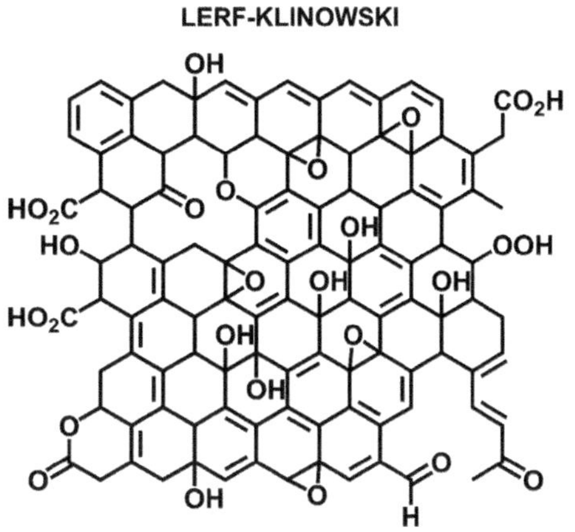

Fig. 2.1 Lerf-Klinowski model of GO structure. Reprinted with permission from [21]. Copyright 2014, Royal Society of Chemistry

and GO by the bright spots and disordered lattice feature [23]. Moreover, high-resolution transmission electron microscopy (HRTEM) is an important characterization method to analyze the atomic level features, including lattice structure, defects, and disordered regions [24].

Additionally, chemical composition can be characterized by Raman spectroscopy [25], X-ray photoelectron spectroscopy (XPS) [26] and Fourier transform infrared spectroscopy (FT-IR) [27]. Raman spectroscopy identifies the degree of oxidation and FTIR can characterize the functional groups, while XPS reveals carbon and oxygen bonds in deconvolution spectra and carbon–oxygen content ratio in the survey spectrum.

For electrical properties, GO is a material with insulating properties, stemming from oxygenated groups on the edge and basal plane of its structure. GO sheets are insulating (sheet resistance: 10^{12} Ω/sq) due to the presence of oxygenated functional groups and disordered lattice structures disrupting electrons pathways in the sp^2 carbon system [28].

Besides electronic properties, GO exhibits unique optical properties, including photoluminescence (PL) in ultraviolet–visible (UV–Vis) and near-infrared (IR) regions, due to its non-zero bandgap unlike graphene. Fluorescence in GO is caused by recombination of electron–hole (e–h) pairs in localized electronic states in sp^2 carbon cluster with sp^3 matrix [29].

In addition, the chemical activity of GO is enhanced compared with pristine graphene. The dispersibility of GO in a polar solvent, including water, acetone, ethanol, Dimethyl sulfoxide (DMSO), dimethylformamide (DMF), n-methyl pyrrolidone (NMP), and Tetrahydrofuran (THF), is enhanced. The oxygen-containing functional groups (e.g., Carboxylic acid) are negatively charged in water, resulting in the repulsive electrostatic interaction between GO larger than van der Waals interaction between graphene layers, which its structure's edge and basal plane prevents GO sheets from aggregation and keeps GO stable in the dispersion [30]. These unique properties allow GO to be used in a variety of engineering applications.

2.2.1 Chemical Synthesis of Graphene Oxide

Methods for GO synthesis are mainly classified into three categories: Brodie method [31], Staudenmaier method [32], and Hummer's method [33]. GO was first produced by Brodie [31] by adding a mixture of potassium chlorate ($KClO_3$) and concentrated nitric acid (HNO_3) to graphite. Staudenmaier's work [32] was developed 40 years later based on Brodie's method; the recipe was altered by reducing the vol of HNO_3 and adding concentrated H_2SO_4. This method increased the acidity of the mixture, dramatically improving the degree of oxidation. However, both methods use $KClO_3$ as the oxidizing agent and HNO_3 as the strong acid, which poses a risk of explosion and toxic gas (NOx and ClO_2) release. In 1958, Hummers and Offeman [33] substituted $KClO_3$ and HNO_3 with potassium permanganate ($KMnO_4$) and sodium nitrate ($NaNO_3$), which negates Cl_2 production. The reaction time of Hummer's method was significantly shortened, and toxicity risks were significantly reduced, compared to Brodie's and Staudenmaier's methods. Due to these advantages, the modified Hummers method has become the most prevalently utilized approach for GO synthesis today.

As GO synthesis methods continue to be developed to avoid using $NaNO_3$ to eliminate the generation of toxic gasses like NO_2 and N_2O_4. Moreover, this method improved oxidation level from 23.99 wt% to 27.97 wt% compared to Staudenmaier's method. Kovtyukhova et al. [34] proposed a pre-oxidation method with phosphorus pentoxide (P_2O_5), potassium persulfate ($K_2S_2O_8$), and H_2SO_4 added to graphite, which could reduce incomplete oxidation. Tour's group [35] demonstrated an improved, simplified synthesis method with reduced toxic gas emission, which entailed mixing concentrated H_2SO_4 and H_3PO_4 in a ratio of 9:1 (by volume) and increasing the amount of $KMnO_4$. However, this method introduced H_3PO_4 as an impurity. Chen et al. [36] found that using only $KMnO_4$ and H_2SO_4 in graphite under a low temperature not only reduced the reaction time but also significantly enhanced the yield of GO. In addition, $KMnO_4$ played the role of strong oxidant and assisted in the complete intercalation of graphite in acidic media, thus negating the necessity of $NaNO_3$ in the production. However, $KMnO_4$ introduced manganese (I) (Mn^+), which polluted the system and appreciably lengthened the purification time. Moreover, as an ingredient in fireworks, $KMnO_4$ poses a high risk of explosion when exposed to temperatures over 55 °C. Thus, Peng et al. [37] substituted $KMnO_4$ with another strong oxidant potassium ferrate (K_2FeO_4) and produced single-layered GO in 1 h. K_2FeO_4 is a green, safe, inexpensive, and effective strong oxidant; using K_2FeO_4 also shortens the reaction time. This method guided a new way to improve Hummer's method in environment-friendly production. For example, Yu et al. [38] proposed an H_2SO_4-free GO synthesis method, using only K_2FeO_4, H_2O_2, and water as precursors. The method is similar to Hummer's method but results in a much higher yield of 65% (Yu's method) compared to 40% (Hummer's method).

Due to its quicker response time and the ease of graphene oxide (GO) production, the Hummers method has become the preferred choice among the three main synthesis methods. Tour's group [39] proposed a three-step theory of GO production from graphite, presented as follows. First, sulfuric acid-graphite intercalation compound (H_2SO_4-GIC) is formed by graphite and acidic media at a low temperature, and oxidation begins from the edges of the graphite and moves towards the middle. Second, the graphite intercalation compound (GIC) transforms to pristine graphite oxide (PGO) with increasing temperature. In this step, $KMnO_4$ starts to oxidize graphite by diffusing $KMnO_4$ into the graphite body, which is an important factor in the determining rate. The reaction rate and the size of the graphite flake also impact the lateral size reduced the reaction time and shape of the PGO. Finally, PGO exfoliation occurs when exposed to water, forming GO. After the GO formation, the solution must be purified with water and weak acid to remove impurities. The second step of the GO production process is a reaction rate-determining step, which can be modulated to control the product. Eigler et al. [40] found that reactions vary within the same batch of GO because small flakes can be oxidized faster than large flakes; high-crystallinity graphite reacts much slower than low-crystallinity one. Chen et al. [41] proposed water-enhanced oxidation to modulate oxygenated groups of GO. At an elevated temperature of 95°C, GO sheets are cut due to oxidation of graphite with an Mn (VII) catalyst. Mn_2O_7 catalyzes water to generate O_3 in the system and improve oxidation; therefore, Mn (VII) plays a significant role in the oxidation process. As a result of oxidation, carboxyl groups are formed as permanent defects in the GO structure, which means this partial structure cannot be restored afterwards. The reaction equations of Mn (VII) with water are shown below.

$$3H_2O + 2Mn(VII) \rightarrow O_3 + 2Mn(IV) \tag{2.1}$$

$$O_3 \overset{Mn(IV)}{\rightarrow} O_2 + \cdot O \tag{2.2}$$

$$O_3 + H_2O \overset{Mn(IV)}{\rightarrow} 2HO\cdot + O_2 \tag{2.3}$$

$$\text{graphite} \overset{(O_3/\cdot O\cdot/HO\cdot)}{\rightarrow} \text{PGO} \tag{2.4}$$

Chen et al. [41] also investigated the destructive oxidation process of GO production under other temperatures. The schematic of the GO synthesis mechanism. They found that under low-temperature synthesis (0 °C) the structural integrity of GO is preserved better than synthesis at a high temperature (95 °C). High temperature caused destructive oxidation to the structure, resulting in permanent defects on the edges. Pendolino et al. [42] drew a similar conclusion when investigating the impact of temperature on GO structure based on Sun's research [43]. They compared the production of PGO at 0°C and 30°C and found that at low temperatures, the exfoliation process dominated to generate PGO, while at high temperatures, bulk structures convert to layered structures. Therefore, there is a tradeoff between structural integrity and layer separation; a temperature balance must be achieved.

2.2.2 Reduction of Graphene Oxide

RGO can be derived from GO by recovering the structure that is changed during the oxidation process in GO synthesis. The electrical conductivity can be described in two ways: sheet resistance (R_s, Ω/sq) and bulk conductivity (σ, S/m). The relation between sheet resistance and bulk conductivity is shown in Eq. 2.5 [44],

$$R_s = \frac{1}{\sigma t} \tag{2.5}$$

where t is the thickness of the sample. The electrical conductivity can be measured by the two-probe or four-point probe method with a microelectronic pair on the RGO film. After GO reducing to RGO, the oxygen-containing functional groups are removed, and delocalized electrons can transport through the recovered graphitic lattice.

Chemical reduction [45] and thermal annealing [46] are the most common conventional ways to reduce GO. However, thermal annealing is disadvantageous because it requires a temperature of up to 1000 °C, which causes unavoidable defects in the structure that reduce the conductivity of RGO. Meanwhile, hydrazine is highly toxic, making it unacceptable for green processes. Therefore, the chemical reduction has been proposed as a simple and economical alternative to these common methods. Reducing agents suggested for the chemical reduction include ascorbic acid (vitamin C) [47], hydrohalic acid [48] (e.g. HI acid), metal hydrides (e.g. sodium borohydride ($NaBH_4$) [49]), and hydrogen [50] (including nascent hydrogen [51]). Ascorbic acid is a non-toxic reductant with excellent water solubility and applicability as a surfactant [47]. In addition to water, ascorbic acid is soluble in *N*, *N*-Dimethylformamide (DMF) and *N*-Methyl-2-Pyrrolidone (NMP), which help establish a stable RGO solution. HI acid is the strongest reductant among hydrohalic acids and is efficient in reducing opening epoxides. The reduction process can be performed on GO powder (solid) or film (gas or solution) at low temperatures [48]. Among metal hybrids, $NaBH_4$ is the most common reducing agent. Although $NaBH_4$ can reduce carboxyl groups, it is unable to completely remove all functional groups, which negatively affects electrical conductivity [49]. The use of hydrogen or nascent hydrogen with catalysts results in an extremely fast and efficient reduction of GO [51]. Compared to pure H_2, the addition of metal pollutes the reaction environment, although metal in acid can produce a steady flow of hydrogen to the system. Although chemical reduction holds advantages over more traditional GO reduction methods, the reductants and reaction processes utilized require further optimization. Table 2.1 presents the different reduction techniques used for RGO synthesis and the respective resistivity and transparency achieved by each process.

Table 2.1 Sheet resistance/conductivity and transparency of RGO

Deposition method	Reduction method	Sheet resistance/ conductivity	Transparency	References
Dip coating	Hydrazine vapor reduction	188 kΩ/sq	98%	[52]
	Thermal annealing	10^4 S/cm	90%	[53]
	Thermal treatment	550 S/cm (1.8 ± 0.08 kΩ/sq)	70%	[54]
	Thermal annealing	60 kΩ/sq	81%	[55]
Electrophoretic deposition	Direct charge transfer	6.4 kΩ/sq	90%	[56]
	Hydrazine + thermal reduction	3.16 Ω/sq (about 500 nm thickness)	–	[57]
Spin coating	Hydrazine vapor reduction	$<10^3$ Ω/sq	>80%	[58]
	Hydrazine-hydrate vapor	10^3 S/m	–	[59, 60]
	Hydrogen with $PdCl_2$	43 Ω/sq (with metal grid)	84%	[50]
		18 Ω/sq (with metal grid)	80%,	
	HI reduction	3.74×10^{-6} Ω · m	70%	[61]
	Thermal annealing	About 800 Ω/sq	82%	[62]
Self-assembly (dip coating)	HI reduction	840 Ω/sq	78%	[63]
	Hydrazine + thermal reduction	11.3 kΩ/sq	87%	[64]
		31.7 kΩ/sq	96%	
Spray coating	Hydrazine + thermal reduction	0.1–0.3 Ω/sq (with AgNWs)	89%	[65]
	Hydrazine monohydrate	2.2 kΩ/sq	84%	[66]
	Hydrazine monohydrate	13.7 Ω/sq (with AgNWs)	–	[67]
Langmuir–Blodgett	Thermal and water vapor treatment	37,720 S/m (nanocomposite)	–	[68]
	HI reduction	1100 Ω/sq (with doping)	91%	[69]

2.2.3 Reduction Mechanism of RGO

The reduction process can be regarded as a two-step S_N2 nucleophilic reaction and a one-step thermal elimination [70]. In the first step, L-AA is oxidized to its first oxidative intermediate, dehydroascorbic acid (DHA), while hydroxy and epoxy groups of GO are attacked by L-AA [70]. Two protons are released from L-AA and have a high binding affinity to these reactive species to form water molecules. Then, the intermediate can be further irreversibly oxidized to guluronic acid and oxalic acid. L-AA and its oxidation products can form hydrogen bonds to interact with residual functional groups, carboxyl groups on the edge of GO [39]. Moreover, the alkaline condition can mitigate π–π stacking by electrostatic repulsion between RGO sheets, preventing aggregation [70]. The S_N2 nucleophilic reaction is reversible and unstable because degradation products are produced at 100 °C at 2 h both in acidic and alkaline conditions. When pH is 10, only small amounts of degradation products are generated but the reactivity of L-AA is decreased. Therefore, not all epoxy and hydroxy groups can be removed. Figure 2.2 shows the chemical structure and reaction process of hydroxy groups and epoxy groups.

The reduction process of GO and HI can be represented as the S_N2 substitution and elimination reaction (see Fig. 2.3). HI can effectively remove epoxy and hydroxy groups. Epoxy groups undergo ring-opening reactions with nucleophiles from HI. Then, iodine is alternative to hydrogen efficiently via S_N2 substitution reaction because the binding energy between carbon and iodine atom is weaker than that between carbon and hydrogen [71]. Therefore, epoxy groups become hydroxy groups. Then, all hydroxy groups in GO can be substituted by iodine via S_N2 substitution reaction and easily eliminated to C=C bonds. With excess HI, however, all hydroxyl groups are substituted by halogen

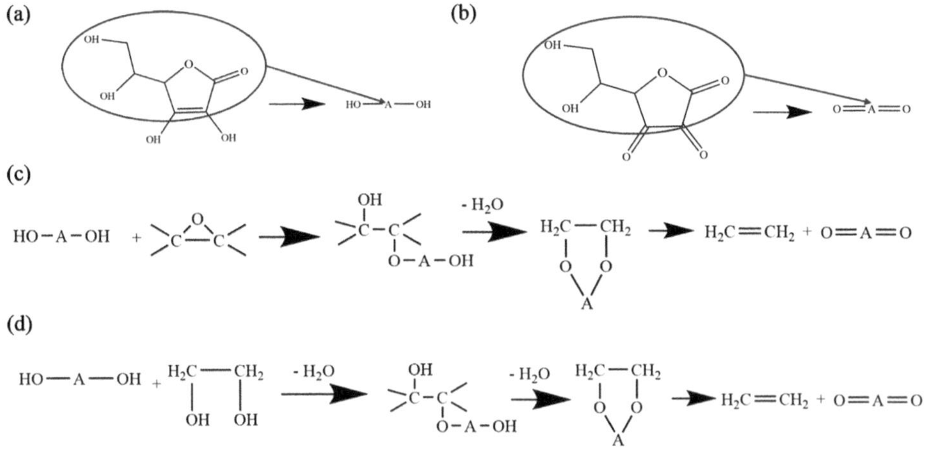

Fig. 2.2 Chemical structure of **a** L-ascorbic acid (L-AA) and **b** Dehydroascorbic acid (DHA); **c** reduction of epoxy groups by L-AA; **d** reduction of hydroxy groups by L-AA [70]

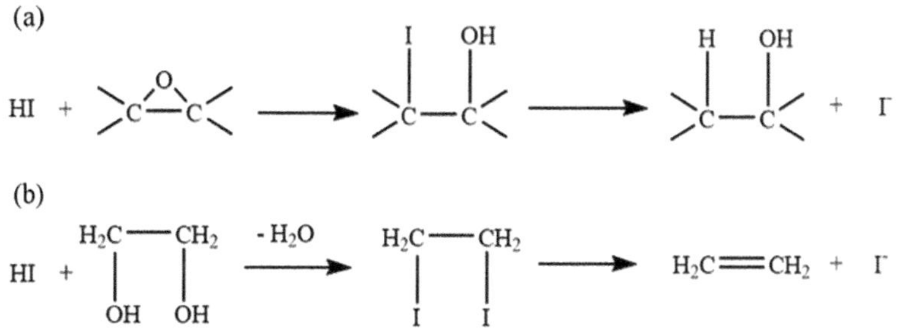

Fig. 2.3 **a** Reduction of epoxy groups by HI and **b** reduction of hydroxy groups by HI [71]

atoms in several steps. The competing reactions between thermal elimination reaction and further substitution influence the degree of lattice restoration. The optimization of reaction time and the amount of HI are necessary to ensure that the elimination reaction is favored over substitution, thereby forming more sp^2 structures. However, not all functional groups can be eliminated. The functional groups attached to the plane and edges have different binding energy. For example, a single hydroxyl group attached to the interior aromatic domain is less stable at room temperature than the one attached to the edge [72]. Therefore, the hydroxy groups attached to the plane are more reactive to be reduced, while the unstable hydroxy groups attached to the plane are dissociated.

Healing the defects in RGO can also recover the electrical conductivity of RGO. According to the second law of thermodynamics, a certain number of defects are present in the crystalline structure [73]. The possible atomic structure of lattice defects is shown in Fig. 2.3. Introduced atoms can change the electron system of sp^2-bonded carbon materials and reconstruct atom arrangements. Point defects, including vacancies or interstitial atoms, are usually present in the lattice structure (Fig. 2.4a). Stone–Wales defects are the unique defects in graphitic structure by reconstruction of the six-ring structure to five or seven-ring (Fig. 2.4c). In addition, defects can come from introduced atoms, carbon atoms and foreign atoms (Fig. 2.4b). The adatom can change the structure arrangement from sp^2-hybridization to the sp^3-bonded structure. The scattering of electron waves at defects has a strong influence on the electrical conductivity and optical properties. The conductivity of graphene is primarily determined by the long-range localized π-electrons of the graphitic lattice [44]. First, bond length change can alter the binding energy between π-electrons and influence the conjugated structure [74]. Furthermore, the binding energy change may influence the reduction process. The difference between carbon and hydrogen and between carbon and iodine may be changed, so the effectiveness of HI is changed. The functional groups cannot be eliminated effectively.

In the reduction process, some defects can be healed to recover the conjugated structure. Hopping and tunnelling interactions between the clusters increased by oxygen

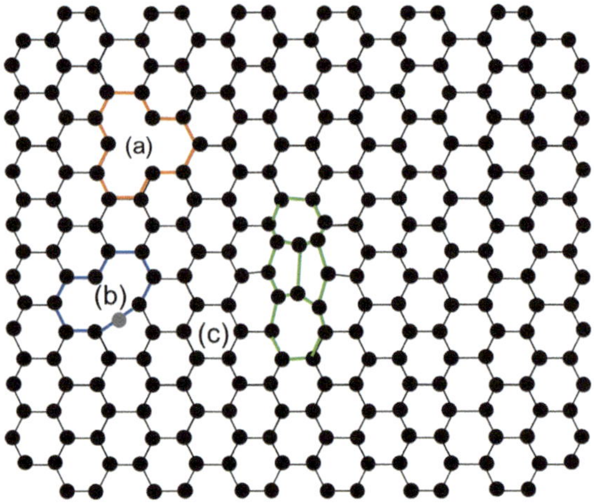

Fig. 2.4 Atomic structure of lattice defects in RGO. **a** Vacancy defects, **b** adatom defects, **c** Stone–Wales (SW) defects [73]

removal result in better connectivity among the original graphitic domains [75]. Theoretically, if the sp^2 structure approaches 60%, the conductivity of GO meets a threshold of percolation [76]. Moreover, in moderate temperature, the functional groups can be removed without introducing new defects, unlike thermal annealing due to migration of adatoms and vacancies expansion increases exponentially with temperature [73].

2.2.4 Work Function of RGO

The work function of RGO is a fundamental surface property. The work function is defined as the minimum energy to extract electrons from the materials. The work function is not an intrinsic property because it is affected by the surface impurities, film uniformity, crystal orientation, and so on. Ultra-violet photoelectron spectroscopy (UPS) is the most used method to measure work function of graphene. To measure the work function, UPS can measure kinetic energy (KE) or binding energy (BE) and the work function can be calculated using Eq. 2.6 [77],

$$\phi = h^{\circ} - (E_{MAX} - E_{MIN}) \tag{2.6}$$

where ϕ is the work function, $h\nu$ is the photon energy of the excitation light source (21.22 eV), E_{MAX} is the energy of the electrons are emitted with kinetic energy h°, and E_{MIN} is the energy of the electrons are emitted at kinetic energy ϕ. For metals, the work function is defined as the energy difference between the Fermi level and vacuum level. However, for semiconductors, Equation $\phi = h\nu - (E_{MAX} - E_{MIN})$# is not the work function but ionization energy (IE) because the electrons are filled in the valence band, not the Fermi level. The work function is calculated using Eq. 2.7 [77],

$$\phi = h^{\circ} - (E_{MAX} - E_{MIN}) - E_{VB} = h\,\nu - (E_F - E_{cutoff}) \quad (2.7)$$

where E_F is fermi level, E_{VB} is valance band, and E_{cutoff} is the secondary electron high binding energy to the cutoff edge.

The work function (WF) can be tuned by changing the factors, related to the structure itself and the application of graphene. The work function depends on the presence of different oxygenated functional groups. Kumar et al. [78] showed functional group concentration changing can vary WF by DFT calculation. The WF of GO increased from 4.4 to 6.8 eV when the C = O ratio was increased from 1.5 to 20%, which was ascribed to the strong dipole moment originating from the C = O double bond, whereas the WF of epoxy groups increased from 4.35 to 5.6 eV, and the WF of hydroxyl groups increased from 4.25 to 4.95 eV [78]. Surface dipoles are caused by the high electronegativity of O atoms on GO, which boosts the WF of GO to 4.9 eV by extracting electrons from graphene [78]. Moreover, defects also change the WF in a wide range. Bae et al. [72] found that graphene with Stone-Thrower-Wales defects decreased the WF, a hydrogen passivated C2 vacancy increased the WF of graphene. Long et al. [79] investigated the effects of grain boundaries and wrinkles on the WF of graphene. The WF increased towards the axial centers but reduced at the edges as it came to wrinkles.

2.3 Graphene Quantum Dots

Graphene quantum dots (GQDs) are the 0D graphene materials with additional exceptional properties as compared to graphene. Due to their size tunability, high optical absorptivity and excellent electrical characteristics, GQDs have captured the attention of researchers; it is currently making appreciable progress in application toward optoelectronics. Furthermore, since GQDs are composed of carbon, one of the most abundant materials on earth, it is a safer and more economical alternative to other most used QDs, which are composed of rare and toxic elements.

2.3.1 Methodologies for Synthesizing GQDs

GQD synthesis methodologies are classified as top-down and bottom-up ones shown in Fig. 2.5. Top-down methods always utilize cheap graphite as precursors and cut bulk material into nanoparticles. The bottom-up method uses molecules to build up nanomaterials. The first GQD solar cell was designed by Yan et al. [80], who employed GQDs as a sensitizer in solar cells using a novel solubilization strategy to design GQD nanostructure. They demonstrated that GQDs can replace traditional sensitizer materials like TiO_2, with the resultant GQD solar cells having comparable short-circuit current, open-circuit voltage and fill factor. Their soluble-chemistry approach to synthesis provided the GQDs

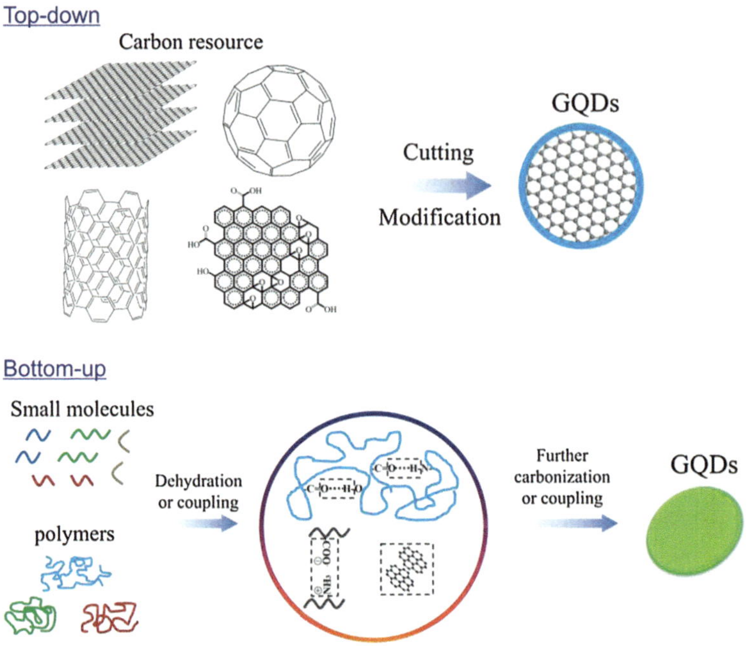

Fig. 2.5 Two categories of GQD synthesis: "Top-down" and "Bottom-up". Reprinted with permission from [81]. Copyright 2017, Elsevier

with high solubility, enabling interactions with other complexes, resulting in improved solar conversion.

Oxidative cleavage is the most frequently used method for GQD synthesis. This is considered a top-down method, in which H_2SO_4, HNO_3, or other acids and oxidizers are applied at high temperatures to cut carbon materials such as coal [82], graphite [83], carbon nanotube [84], black carbon [85], and carbon fibers [86], which are then formed into GQDs. The first top-down method of GQD synthesis was demonstrated by Pan et al. [87], who synthesized strong, blue-emitting GQDs by hydrothermal treatment of graphene sheet (GS) using concentrated H_2SO_4 and HNO_3. Shen et al. [88] used GO sheets that were cleaved into small pieces using HNO_3, then surface passivated via the addition of ethylene glycol. They then added hydrazine hydrate to reduce GO, obtaining GQDs. Their ultrafine, blue-emitting GQDs were utilized as a new fluorescent material, applicable as an up-converter with considerable potential for application in the energy conversion industry, biotechnology, and so on. However, the use of strong oxidizers in oxidative cleavage can be dangerous, as reactions with strong oxidizers may cause explosions. Therefore, similar but less volatile hydrothermal and solvothermal methods have been adopted for GQD synthesis. Tian et al. [89] developed a one-step solvothermal method using a graphite precursor and hydrogen peroxide in a DMF environment; unlike methods that utilize traditional oxidizers such as H_2SO_4 and HNO_3, no impurities were produced in the entirety of this synthesis procedure. The resultant GQDs showed a 15% quantum yield under different pH conditions, with excellent PL stability. Hydrothermal and solvothermal methods emphasize simplicity, allowing scalability of the technique; hydrothermal methods are environmentally friendly. However, such techniques are disadvantageously associated with high temperature, high pressure, and long reaction times.

Bottom-up methods for GQD synthesis entail aggregating small molecules to form large GQDs, with precise control over GQD size, shape, and total number of carbon atoms [81]. Drawbacks of bottom-up methods include difficulties in the multi-step fabrication process and low production yield. The most common method for producing GQDs is cleaving carbon-based materials by oxidation and reduction. As researchers continue to investigate new techniques to produce GQDs, requirements of environmental compatibility, high quantum yield, and high GQD qualities are the main focus.

2.3.2 Optical Properties of GQDs

GQDs exhibit similar optical properties with GO in UV–Vis and photoluminescence (PL) spectra. In the UV–Vis spectrum, the main peak at 230–270 nm features π–π^* transition of $C = C$ bonds and a shoulder peak at around 320 nm with a long tail is the result of n–π^* transition of $C = O$ bonds [90]. For PL, GQD can emit different colors, mainly from yellow to blue, depending on the size and chemical compositions in different synthesis methods [91]. The emission spectra of GQDs are broad, and is controlled by the core and the surrounding chemical groups. PL intrinsic center is determined by the conjugated π-domains, which results from the quantum confinement effects (QCE), which is defined as observed when the size of the QD is too small to be comparable to the Bohr radius [92]. The absorption peak energy falls as the size of GQDs increases. The surrounding chemical groups of GQDs control the surface state, especially prepared by "top-down" methods. Zhu et al. found green emission is mainly attributed to carboxyl and amid groups and blue color is determined by hydroxyl groups [93]. Moreover, defects including edge shapes attached to functional groups and doping heteroatoms influence the PL properties of GQDs. Pan et al. proved that zigzag sites with a carbene/carbyne-like triplet ground state $\sigma^1\pi^1$ might contribute to blue emission [94]. Tetsuka et al. developed the amino-functionalized GQDs with wide tunability of narrow PL and the electronic structure was controlled by their structure terminated by the edge with primary amine [95]. Jin et al. functionalized GQDs with amine groups by a two-step cutting process from GO and observed a redshift of PL emission compared to unfunctionalized GQDs [96]. Therefore, the optical properties of GQD are complicated due to various factors tailoring the GQDs chemical structure.

2.4 Potential Applications of Graphene and Its Derivatives in Device Engineering

2.4.1 MOS Devices

Figure 2.6 shows the typical MOS capacitors (MOSCAP), which are made of an oxide layer between the metal gate and the semiconductor body. The semiconductor is connected to a back contact. The oxide layer functions as the dielectric layer, and the area of metal is the range of the capacitor. In MOS devices, the MOSCAP is a part of the completed electronics devices metal–oxide–semiconductor-field-effect-transistors (MOSFETs), which have source and drain more than MOSCAP. Capacitance–voltage (C–V) characteristics are commonly used to study the quality of MOS devices, including oxide thicknesses, threshold voltage, flat-band voltage, and oxide quality. The general way to measure the C-V curve is by applying DC sweep voltage and a small AC sinusoidal signal (1 kHz–10 MHz). As the sweeping in no voltage, negative voltage and positive voltage applied, which is gate voltage (V_g) compared with flat band voltage (V_{fb}), the devices pass through accumulation, depletion, and inversion regions, respectively. In the accumulation region, negative and positive charges are in equilibrium at the interface between the semiconductor layer and the oxide layer. The oxide capacitance (C_{ox}) can be measured in the high-frequency strong accumulation flat curve. In this situation, the MOSCAP is a parallel-plate capacitor model, and the oxide thickness (t_{ox}) can be calculated from Eq. 2.8,

$$C_{ox} = \frac{A\varepsilon_{ox}}{t_{ox}} \tag{2.8}$$

where A is the gate metal area, ε_{ox} is the permittivity of the oxide material. In the depletion region, the majority carries are depleted so the MOSCAP is an insulator. As the voltage increase, the depletion region moves far away from the gate, so the effective oxide thickness increases and capacitance decreases. In the inversion region, the electrons accumulate at the oxide–semiconductor interface, which is inverted in the carrier polarity. As the gate voltage is above the threshold voltage, most electrons are in the inversion layer, so the C-V is the flat curve.

Graphene can be used as the top contact in MOS devices. In 2008, Lemme et al. [97] fabricated the first graphene field-effect device by depositing mono- or few-layer graphene on SiO_2 silicon wafers as a top gate. The devices confirmed that graphene can achieve field effects and exhibited two advantages: (1) The carrier mobilities of this device are higher than in silicon; and (2) The top gate can modulate the drain current by the top gate field effects [97]. However, the band-gap tunning, and work function mismatch are the problems to be solved. Park et al. [98] applied monolayer graphene as a gate electrode on top of three insulating layers in a charge trap flash (CTF) memory device. The high work function of p-type graphene reduced the trap-assisted tunnelling (TAT) current at the gate

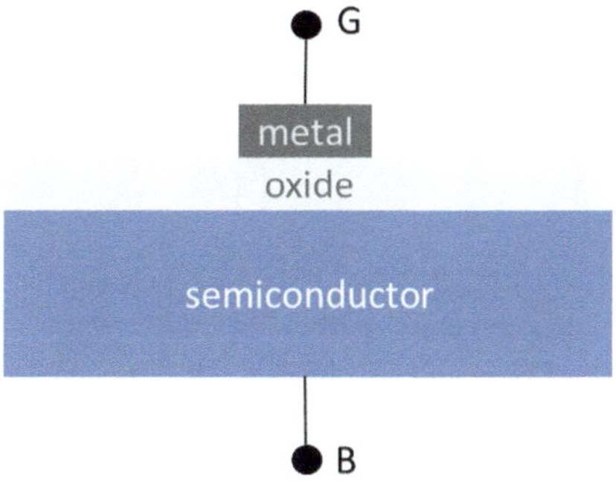

Fig. 2.6 The scheme of the MOSCAP structure

electrode [98]. This device improved the quality of gate dielectric due to the presence of graphene and represented the new era of graphene applied in high-performance electronic devices based on MOS structure.

2.4.2 Graphene-Based Materials as Transparent Conductive Electrodes (TCEs)

Transparent conductive electrodes (TCEs) play an important role in optoelectronic devices such as touch screens, displays and solar cells. Currently, indium tin oxide (ITO) is the most used TCE in industry and research due to its high conductivity and transparency. However, ITO lacks flexibility and is costly due to the scarcity of indium. Thus, alternative materials are under investigation, with carbon nanomaterials being considered the best choice to replace ITO [58]. Transparency of single-layer graphene can reach up to 97%, while sheet resistance can reach up to hundreds of Ω/sq. Graphene-based materials (RGO, few-layered graphene, graphene nanocomposites) with low sheet resistance and high transparency prove a low-cost option to replace ITO as a TCE in solar cells. Wang et al. [54] proposed the first RGO transparent electrode in a dye-sensitized solar cell (DSSC) (Fig. 2.7a); the RGO electrode displayed a sheet resistance of $1.8 \pm 0.08 k\Omega/sq$ and more than 70% transparency from 1000 to 3000 nm; the DSSC's power conversion efficiency was measured to be 0.26%. Ricciardulli et al. [67] developed an AgNWs–exfoliated graphene (EG) TCO, which was used as a bottom electrode in OSC and polymer LED (PLED), with PCE = 6.6% and EQE = 4.4% respectively (Fig. 2.7b, c). This research proved that 1D-2D electrodes can achieve success in bulk heterojunction (BHJ) OSCs.

The key properties of TCOs used in solar cells include (1) low sheet resistance (<3 kΩ/sq); (2) high transparency (>70%); (3) uniform film; (4) excellent chemical and mechanical stability. For graphene-based materials, the challenge lies in achieving low sheet resistance. RGO is produced from GO by restoring the structure of graphene, which cannot avoid the creation of defects in the structure, resulting in disrupting the charge transfer

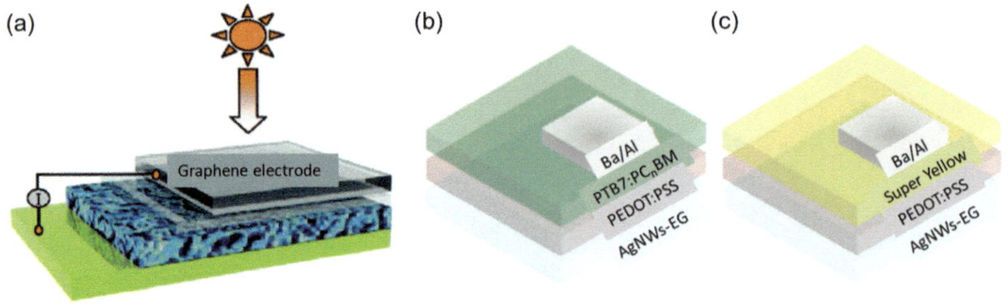

Fig. 2.7 **a** Schematic of DSSC with graphene-based TCE. The layers from top to bottom are graphene anode, TiO_2 (ETL), dye, spiro-OMeTAD (HTL), Au (cathode). Reprinted with permission from [54]. Copyright 2008, ACS publications. Schematic of **b** OSC (Ba-Al/PTB7:$PC_{71}BM$/PEDOT:PSS/AgNWs-EG) and **c** PLED structure (Ba-Al/Super Yellow/PEDOT:PSS/AgNWs-EG). Reprinted with permission from [67]. Copyright 2018, Wiley

process, and thereby decreasing electrical conductivity. By increasing the number of layers, the charge transfer channel can be tailored to increase conductivity, with the caveat of decreased transparency due to increased thickness. Therefore, establishing a trade-off between conductivity and transparency of graphene-based TCEs is essential.

2.4.3 Role of Graphene-Based Materials in Solar Cell Active Layers

Graphene-based materials (e.g. RGO, graphene) can be used as the active layer in solar cells. RGO and graphene with high charge mobility can be functionalized as electron transport layer (ETL) and hole transport layer (HTL). Yang et al. [99] incorporated RGO as a 2D bridge into TiO_2 electrode, resulting in enhanced electron transport mobility and conductivity compared to CNT-TiO_2 electrode. A schematic representing the electrode structures and the operating principle is depicted in Fig. 2.8a–f. Apart from ETL, graphene and its derivatives can also be applied as HTL. Wu et al. [100] employed GO as HTL and $CH_3NH_3PBI_{3-x}Cl_x$ as an absorber in an inverted planar heterojunction (PHJ) perovskite solar cell (PSC) (Fig. 2.8g, h, achieving a PCE of 12.4% with a work function of around 4.9 eV. As a hole conductor, GO can extract holes from perovskite, which was confirmed by photoluminescence (PL) quenching efficiency. PL quenching efficiency increases with increasing thickness of GO thin film, indicating a change transfer process between GO and perovskite.

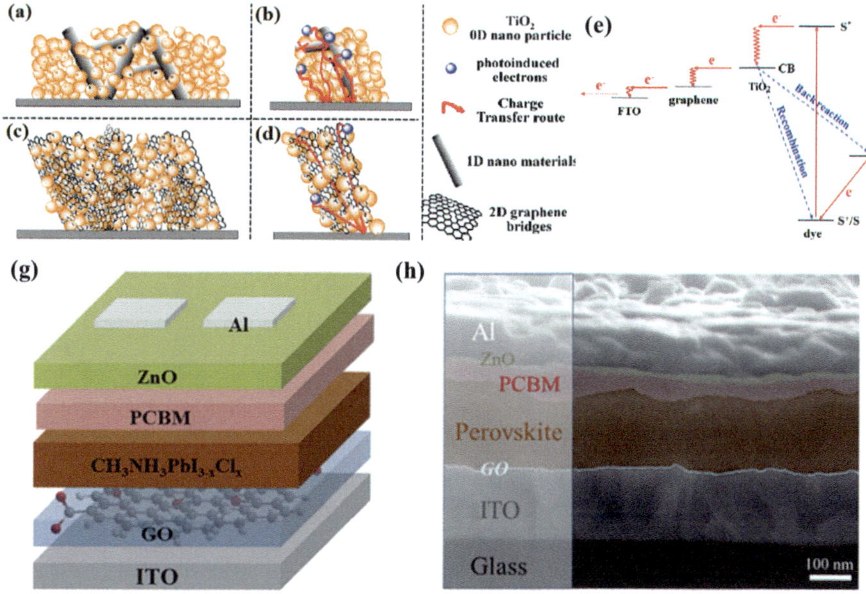

Fig. 2.8 Scheme of **a**, **c** 1D CNT-TiO_2 and **b**, **d** 2D RGO-TiO_2 composite electrode structure **f** operation principle of the device. Reprinted with permission from [99]. Copyright 2010, ACS Publications, **g** Structure of ZnO/PCBM/$CH_3NH_3PBI_{3-x}Cl_x$/GO/ITO PSC and **h** SEM image of device structure. Reprinted with permission from [100]. Copyright 2014, Royal Society of Chemistry

2.4.4 Light Absorption in Solar Cells via Graphene Quantum Dots

GQDs are the ideal candidates for the energy-down-shift (EDS) layer in photovoltaics. Lee et al. [101] synthesized GQDs and spray-coated the synthesized GQDs onto silicon solar cells. A red-shift of PL peaks confirmed EDS and increased efficiency. Figure 2.9 represents the structure of a silicon solar cell with GQD EDS spray-coated on it. QDs are an excellent choice for EDS, as they can be successfully incorporated into other complete photovoltaic devices such as silicon solar cells [101] and perovskite solar cells [102] by spray coating. However, EDS should meet the requirements of high transparency and large Stokes shift exhibiting higher optical efficiency. QD films are not always transparent and after spray coating, may inadvertently block the path of incident light into the device, resulting in decreased efficiency.

2.4.5 Graphene Quantum Dots-Based Photodetectors

The inability of long-wavelength photons to contribute to the generation of electron–hole pairs is the fundamental limit of photodetectors. The bandgap of graphene can be tuned

Fig. 2.9 Demonstration of GQD EDS spray-coated silicon solar cell. Reprinted with permission from [101]. Copyright 2015, ACS Publications

for the required spectral range. In photodetectors, graphene can be used to replace metal in the electrode. Additionally, by tuning its bandgap and combining it with other semiconductors that have strong UV absorption (e.g. ZnO), it is utilizable as a UV-sensitive material. Zhang et al. [103] fabricated a solution-processed GQD DUV photodetector with high sensitivity and fast response time to short-wavelength (254 nm) (Fig. 2.10a). Due to quantum confinement, the GQD bandgap can be modulated to 3.8 eV. With such a wide bandgap, photodetectors can exhibit high responsivity in the DUV range. Graphene integrated with PbS QDs in photodetectors exhibited ultrahigh responsivity (8.61 A/w) and detectivity (7×10^{13} J) [103].

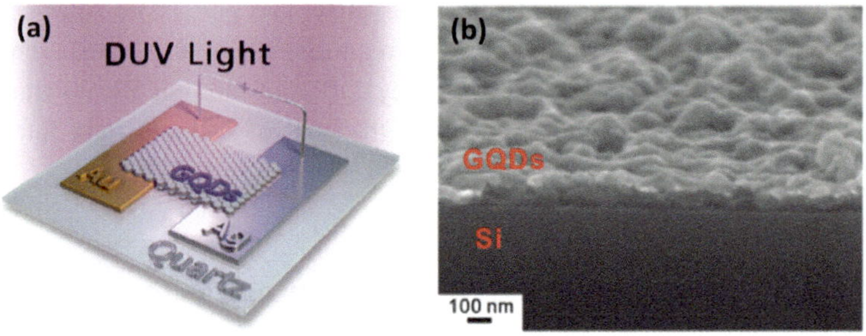

Fig. 2.10 **a** Scheme of the GQD DUV photodetector with Ag-Au electrodes. **b** SEM image of GQDs and Si substrate. Reprinted with permission from [103]. Copyright 2015, ACS Publications

3 Synthesis of Graphene Oxide for Solar Cell Applications

3.1 Synthesis and Characterization of Graphene Oxide

GO was prepared based on Hummer's method. 1.0 g of natural graphite flakes (200 mesh, Sigma-Aldrich, 75μ > 70%) was oxidized by $KMnO_4$ (MilliporeSigma, ACS reagent, ≥99.0%) and concentrated H_2SO_4 (Fisherbrand, ACS-PUR). Typically, 1 g of graphite flakes was added into 30 mL of concentrated H_2SO_4 in a 100 mL round-bottom flask cooled with an ice bath. The mixture was stirred at 350 rpm until the suspension temperature reached 0 °C. The suspension was vigorously agitated for 30 min, then 3.15 g (3 g + 5% excess) $KMnO_4$ was slowly added (over 20 min or so). The stirring speed was subsequently increased to 500 rpm and maintained for 10 min until the temperature reached below 10 °C. The flask was then removed from the ice bath and placed in a 35 °C water bath with a stir speed of 600 rpm for 3 h to finish the reactions. After that, the mixture was mixed with 300 ml of water in a 1000-ml beaker, followed by the addition of a small amount of 30% H_2O_2 to remove MnO_4^- and MnO_2. The mixture was placed in an ice bath, and then it turned yellow, with small gold pieces suspended inside.

The produced GO dispersion was purified by centrifugation. The water in the suspension was removed by centrifuging at 4,000 rpm for 10 min. Then, 120 mL of 3% HCl was added, and metal ions were removed by sonicating the mixture for 5 min. After that, the diluted HCl was removed by successively centrifuging at 4,000 rpm for 10 min. Finally, the GO dispersion was centrifuged with 3 L water to achieve a pH of 5 or so. The GO dispersion was stored in a sealed glass beaker, wrapped with Aluminium foil, and placed in a dark environment to avoid degradation from light exposure.

F. Sun, *Synthesis and Device Applications of Graphene Derivatives and Quantum Dots*,
Synthesis Lectures on Green Energy and Technology,
https://doi.org/10.1007/978-3-031-56403-1_3

3.1.1 Materials Characterizations

To study the morphology of GO film, the samples were characterized by a Scanning electron microscope (SEM) equipped with Energy-dispersive X-ray spectroscopy (EDS) (SEM/EDS, Leo 1530, Carl Zeiss AG, German) and High-resolution transmission electron microscopy (HRTEM, Libra 200, Carl Zeiss Microscopy AG, German). The sample was prepared by drop-casting onto an RCA-cleaned silicon substrate until the water completely evaporates. SEM images were obtained under a low acceleration voltage of 10.00 kV and the HRTEM was performed to obtain TEM and HRTEM images. The TEM samples were made by dropping diluted GO dispersion onto a lacey carbon TEM grid. The TEM image was obtained using TEM mode with a high acceleration voltage of 200 kV.

Raman spectroscopy (Ramascope dual-wavelength micro-Raman spectrometer, Renishaw, Canada) was used to determine the structure of GO equipped with a reflected light microscope, with a Raman shift resolution of 0.5 wavenumbers. Two laser wavelengths were used for excitation: 488 nm (blue, Ar ion) and 633 nm (red, He–Ne). The sample was prepared by spin coating concentrated GO dispersion onto an RCA-cleaned silicon substrate. The Raman spectroscopy images were taken with a 633 nm excitation wavelength.

Atomic force microscopy (AFM) (Nanoman VS DI D3500, Veeco, USA) was used to measure lateral size, thickness, and layers of GO. The sample was prepared by spin-coating diluted GO dispersion onto an RCA-cleaned silicon substrate. The AFM was operated with tapping mode to measure the GO surface and determine the height of the GO layers.

Fourier transform infrared spectroscopy (FTIR) (Vertex 70, Bruker, USA) was used to determine the functional groups. The sample was prepared by drop-casting concentrated GO dispersion on a KBr slide and drying it using a desiccator.

Photoluminescence (PL) spectroscopy (Edinburg Instruments, FLS920, USA) was used to study the emission spectra of GQD, under excitation at 380 nm with a 900-W Xenon lamp as excitation source, at room temperature. A 395 nm long-pass filter was used to differentiate the second-order resonance peaks of excitation. The excitation spectrum was under 583 nm emission.

X-ray photoelectron spectroscopy (XPS) (VG ESCALab 250, Thermo Fisher Scientific, USA) was used to characterize the chemical composition, chemical state, and molecular structure of the sample surface, usually for approximately 10 nm for polymers [104]. The samples were deposited on the RCA-cleaned silicon substrate by drop-casting and drying in the desiccator. Before being measured by XPS, plasma cleaning the samples by sputtering removed the contaminants layer. Moreover, for insulating materials, the emission of electrons accumulates the positive charge on the surface, so a compensation process was applied to the samples to neutralize the charge on the surface by refilling the electrons from a combination ion gun. The source is monochromated Al K-alpha as the x-ray source.

3.1.2 Measurement of Yield

The yield of GO (Y_{GO}) is calculated using Eq. 3.1

$$Y_{\mathrm{GO}} = \frac{C_{GO} V_{GO}}{m_{Gr}} \times 100\% \tag{3.1}$$

where C_{GO} is the concentration of purified GO (in mg/mL), V_{GO} is the volume of the GO dispersion (in mL), and m_{Gr} is the weight of original material graphite (in mg). The mass of GO was determined by measuring the weight difference between the empty container, and the concentration of GO is calculated by the mass of GO divided by the container with a certain volume of purified GO dispersion. Using the above equation, the yield was calculated to be approximately 50%.

3.2 Results and Discussion for Materials Characterizations

3.2.1 Morphological and Structural Characterizations

Figure 3.1 depicts the morphology and structure of dry GO under different magnifications. Figure 3.1a shows the edge of one piece of GO with many wrinkled layers. Figure 3.1b shows the square-shaped GO sheets, which are single-layered, few-layered, or multiple-layered GO stacked together, with different thicknesses and shapes. Some of the large GO sheets have an area in the range of hundreds of square micrometres (Fig. 3.1c), indicating great potential for high conductivity. The GO sheets have a layered structure, forming ultrathin homogeneous films. Some are folded and wrinkled on the edges, distinguishing them from individual GO sheets.

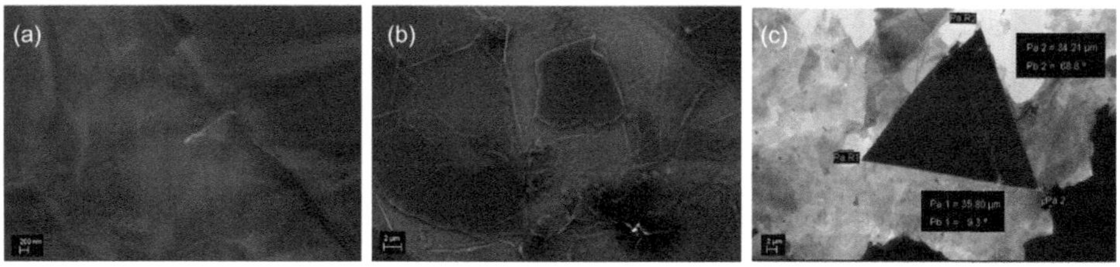

Fig. 3.1 SEM images of GO **a** edge of GO sheet, **b** regular shaped GO sheets, **c** a large area (115.8 2 μm^2) GO sheet

3.2.2 Crystallinity Analysis

Figure 3.2 shows the HRTEM images and Raman spectra, which indicate the crystallinity of the GO structure. The HRTEM images, including a SAED pattern of GO at different magnifications and the Raman spectra, show the structure of GO at the atomic level. The multi-layered structure of GO sheets can be clearly observed in Fig. 3.2a. Moreover, the molecular level of the honeycomb structure can be seen. Several carbon hexagons combine to form the honeycomb structure, indicating that the structure is intact basal plane. The surface of GO is smooth across the entire area of the GO sheet, even where functional groups have been inserted into layers of graphite. In some areas, the carbon lattice is rearranged or vacant, such as Stone–Wales defects. The structure has few defects, showing it to be an excellent precursor for RGO synthesis and further functionalization. SAED pattern is the reciprocal lattice of GO, which can be used to understand crystal structure (Fig. 3.2b). The dark circular shape in the middle arises from the probe. There are three concentric rings with different shades, extending from the middle. Each ring represents each layer of GO. The white specks surrounding the rings represent lattice points. These images indicate that this is a multiple-layer structure GO sheet with incommensurable stacking of hexagonal structures, which results from functional groups interacting with carbon [105].

Raman measurements in Fig. 3.2c were taken to investigate the physical properties of GO, including exfoliation degree and defects [106]. For the Raman spectrum of single-layer graphene, the intensity band assigned for G represented for primary vibrational mode in the single crystal is at around 1580 cm^{-1}, D band standing for structural defects and edges appears in crystallites at 1,300–1,350 cm^{-1} [107]. The structural imperfections mainly arise from the attachment of the hydroxyl group and epoxide group on the basal plane. In addition, small broadbands, called 2D peak or G' peak, appear at around 2,700 cm^{-1} under 632.8 nm laser excitation [108]. Figure 3.2c also shows several peaks: the D peak at 1,326 cm^{-1}; the G peak at 1,584 cm^{-1}; and the broad 2D peak contributed by two peaks at 2627 cm^{-1} and 2917 cm^{-1}. These peaks confirm the production

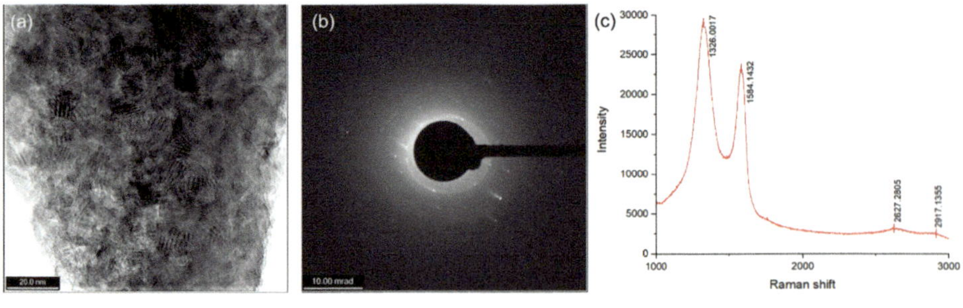

Fig. 3.2 **a** HRTEM image of GO, **b** selected area diffraction (SAED) pattern, **c** Raman spectrum of GO

of GO. They also indicate defects in the structure induced by oxygenated groups. With abundant functional groups, defects (represented by D peak) increase greatly. However, these defects are not particularly concerning because they are not permanent and can be rectified after the reduction of GO to RGO.

The ratio of the intensity of the D peak to the G peak, I_D/I_G, is 1.25. This intensity can be used to measure defects present in carbon materials. Since the D peak shows sp^3 bonded carbon atoms by out-of-plane vibration and the G peak shows sp^2 bonding by vibrational mode, I_D/I_G indicates the sp^3/sp^2 carbon ratio, with larger values associated with highly oxidized materials [109]. For ideal single-layered graphene, there is no D peak. Moreover, I_D/I_G can also show the in-plane crystallite size, L_a, which can be calculated using Eq. 3.2 [108]:

$$L_{\mathrm{a}}(nm) = \left(2.4 \times 10^{-10}\right)\lambda_I^4\left(\frac{I_D}{I_G}\right)^{-1} \tag{3.2}$$

where λ_I (nm) is the wavelength of the laser source, which impacts I_D/I_G. L_a is calculated to be 30 nm or so.

2D bands split into wider bands with high frequency as functional groups are added into the system, attributed to increased interaction between layers [110]. The split peak positions can be used as a reference for several layers. The single peak is at $2{,}678 \pm 1\ \mathrm{cm}^{-1}$ [111]. As the number of layers increases, the peaks split and change positions. The value for the splitting is around 290 cm^{-1}, indicating that most sheets are multiple layers. Intensity of 2D to G peak (I_{2D}/I_G) roughly shows the number of layers. As I_{2D}/I_G is smaller than 0.5, it indicates multiple layers ($n > 10$) [112].

3.2.3 Photoluminescence Analysis

Figure 3.3 shows the PL emission and excitation spectra of GO. Figure 3.3a shows that two emission peaks in the emission spectrum of GO are at 435 nm and 496 nm under 380 nm laser excitation. Figure 3.3b shows the excitation spectrum of GO with the excitation peak at 275.5 nm. GO displays PL emission in the visible range with two peaks; one is a blue band, and another is a long wavelength (LW) band. The dominance of the peaks is highly dependent on pH. The GO dispersion in water has a pH of 6 or so measured by pH paper, such that the LW band is dominant, corresponding to the optical transition from the carboxylic group to carboxyl acid [113]. Therefore, the PL spectra proves the presence of oxygen-containing functional groups. The types of functional groups are analyzed by FTIR and XPS.

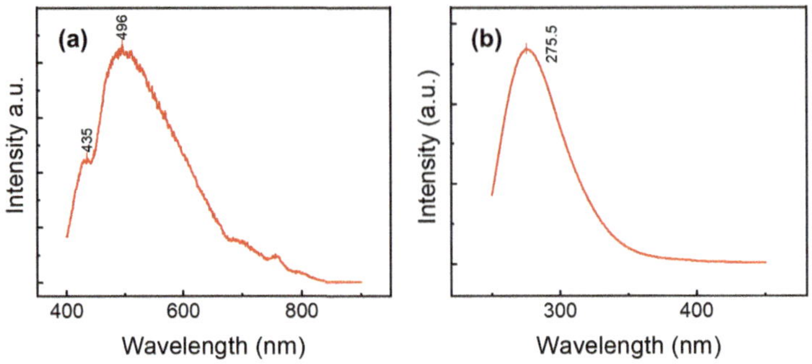

Fig. 3.3 PL spectra of GO; **a** emission spectrum, **b** excitation spectrum

3.2.4 Chemical Composition Analysis

Figure 3.4 shows the FTIR and XPS spectrum of GO to analyze the chemical composition. XPS is used to characterize the functional groups of the sample surface. The different oxygenated functional groups can be characterized in C 1 s spectrum. Figure 3.4a shows the peaks at 283.6 eV, 285.7 eV, and 286.3 eV, corresponding to carbon–carbon double bonds (C=C), epoxy groups (C–O–C), and carboxyl groups (–COOH), respectively [114]. Figure 3.4b shows the carbon peak is at 286 eV and the oxygen peak is at 531.5 eV. The calculated intensity of the carbon–oxygen atomic mass ratio (C/O) is 0.56. The high I_D/I_G ratio indicates sp^3-hybridization caused by high degree of oxidation, which coordinates with low C/O ratio in XPS result. In Fig. 3.4c, the FTIR spectrum shows different types of functional groups at various degrees. The most important feature of GO is the O–H stretching vibration at around 3,000–3,500 cm^{-1} [115]. The C = O, C–C, and C–O–C stretching are at around 1,730, 1,620 and 1,250 cm^{-1}, respectively [116]. The polar functional groups, especially hydroxyl groups attached to the surface resulting in the form of hydrogen bonds with water molecules and presenting high hydrophilicity properties.

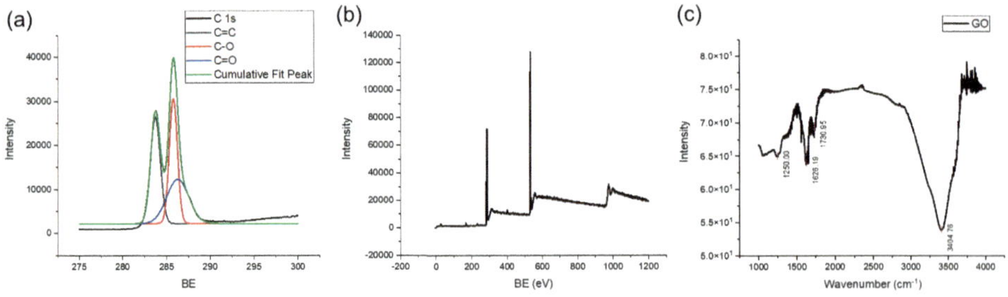

Fig. 3.4 **a** Deconvolution spectrum of C 1 s, **b** XPS survey spectrum and **c** FTIR spectrum of GO

3.3 Summary

In summary, GO was prepared using the modified Hummer's method. The synthesized GO with high yield has different functional groups attached to the surface and edge. The large size of the GO sheet is possible to achieve high-quality RGO in the reduction process. The wrinkles and defects on GO are due to the oxidation and expansion of graphite, which can be partially restored in the reduction process. This environmentally friendly method shows greatly preparing to GO in mass production simply and cost-effectively.

From Graphene Oxide to Reduced Graphene Oxide

4

4.1 Experimental Process of Reduced Graphene Oxide

4.1.1 Formation of GO Thin Film

The as-synthesized GO was diluted to 1 mg/mL and sonicated for 10 min to form a homogeneous solution. Then, the aqueous GO dispersion was deposited on an RCA cleaned glass substrate by spin coating. Spin coating was optimized by starting at 500 rpm for 10 s to initialize the spreading of the solution on the substrate, then increasing to 1000 rpm for 30 s to thin the solution layer, and finally at 250 rpm for 25 s to dry the film while maintaining thickness [118].

4.1.2 Green Reduction of GO: Ascorbic Acid and HI Methods

The GO material produced as described above was used in all the reduction experiments to synthesize RGO. Two reducing agents, L-ascorbic acid (L-AA), and hydrogen iodide (HI) were employed individually and successively. 10 ml of L-AA solution with a concentration of 3.5 mg/ml was sonicated for 10 min to form a uniform mixture. An RCA-cleaned glass substrate coated with GO thin film was immersed into the L-AA solution in a glass petri dish. In the two-step process, the first reduction step reduces GO to an intermediate RGO, which is then further reduced to the final RGO. Several experiments were carried out with L-AA (99%) and HI (distilled, 57 wt.% in H_2O) in different combinations (L-AA only, HI only, HI vapor, HI vapor + L-AA, L-AA + HI, and HI + L-AA) with the HI reductant used in both liquid and vapor forms. The reduction temperature was 90°C for with a duration in the range of 18–24 h, except for the HI vapor (5 h) case in which the GO film on glass was exposed to vapor produced from boiling HI.

F. Sun, *Synthesis and Device Applications of Graphene Derivatives and Quantum Dots*,
Synthesis Lectures on Green Energy and Technology,
https://doi.org/10.1007/978-3-031-56403-1_4

Table 4.1 Two-step reduction to the synthesis of transparent RGO films

First reduction			Second reduction		
Reductant	Temp., time	Intermediate RGO film	Reductant	Temp., time	Final RGO film
HI solution	90 °C, 15 h	RGO_{HI}	L-AA solution	90 °C, 15 h	$RGO_{HI + L\text{-}AA}$
HI vapor	150 °C, 5 h	$RGO_{HI\ vapor}$	L-AA solution	90 °C, 15 h	$RGO_{HI\ vapor + L\text{-}AA}$
L-AA solution	90 °C, 15 h	$RGO_{L\text{-}AA}$	HI solution	90 °C, 15 h	$RGO_{L\text{-}AA + HI}$

Depending on the combination, the intermediate ($RGO_{L\text{-}AA}$, RGO_{HI}, RGO_{HI} vapor) and final ($RGO_{L\text{-}AA + HI}$, $RGO_{HI + L\text{-}AA}$, $RGO_{HI\ vapor + L\text{-}AA}$,) RGO materials were obtained. A two-step reduction process was performed by coupling reducing agents, HI and L-AA, to reduce GO to RGO films in Table 4.1. In the first reduction step, the GO film is reduced to the intermediate RGO film, which is then reduced to the final RGO film in the second reduction step.

4.1.3 Fabrication of Free-Standing RGO Films

The as-prepared GO solution was sonicated for 10 min to form a homogeneous solution. 2 ml of GO solution was deposited on a Whatman® Anodisc inorganic filter membrane (diameter: 13 mm; pore size: 0.02 μm). Then, the GO solution was dried at room temperature by desiccant overnight. After drying, the GO solution became a paper-like film, which can be peeled off, on the Anodisc membrane. The free-standing GO paper was immersed into L-AA and then HI for 48 h, heated on a hot plate at 150 °C. The produced RGO film is named A-RGO.

The as-prepared GO solution was sonicated for 10 min. 10 ml of the GO solution in a glass bottle was heated at 150 °C by the hotplate until the solvent completely evaporated. Then the GO solution became a paper-like film attached to the wall of the bottle. Then, the film was kept annealed on the hot plate at 150 °C for 48 h. The produced RGO film is named S-RGO.

The key to making a free-standing thin film is to separate the thin film and the substrate without damaging the film. Unlike glass or silicon substrate, Anodisc membrane filter is made of alumina matrix with honeycomb pore to decrease the interaction between thin film and substrate. During the thermal annealing to reduce GO to RGO, the thin film becomes rigid, this may be due to the decreased layer distance so the increased interaction between RGO sheets larger than the interaction between RGO sheets and the substrate. Therefore, after thermal annealing, the RGO thin film can be peeled off.

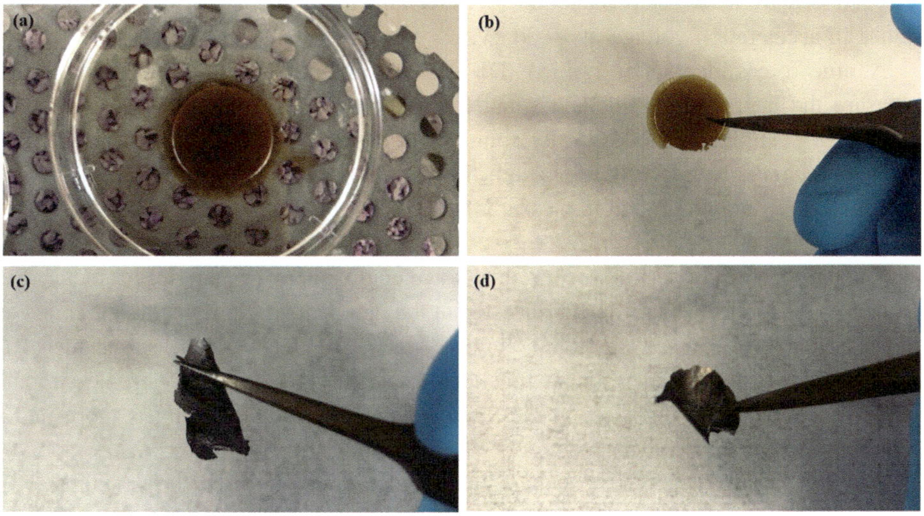

Fig. 4.1 Optical images of **a** graphene oxide (GO) thin-film before peeled off; **b** GO free-standing thin film (A-GO); **c** reduced graphene oxide (RGO) free-standing thin film drying through Anodisc membrane filter (A-RGO); **d** RGO free-standing thin film by water evaporation (S-RGO)

Figure 4.1 shows the optical images of produced GO and RGO free-standing thin film. The produced RGO thin film is flexible and foldable. When it is folded, the RGO paper is not broken. Similarly, when the water in the suspension evaporates, the concentration of GO in the suspension rises, resulting in a large increase in sheet-to-sheet contacts and eventually stacking the sheets on top of each other in the deposition process, generating a layer-by-layer nanostructure [119]. The self-assembly of the GO thin film is formed in the liquid–air interface by heat aggregation, which is driven by Brownian motion, and the hydrosol is decreased [120]. Therefore, the GO sheets have more opportunities to collide and aggregate with each other when water is evaporated from the hydrosol and new layers are aggregated with the existing layers in the liquid/air interface by van der Waals interaction [120]. When the evaporation process is finished, GO thin film is formed to hang in the air with edges attached to the bottle wall. Then, the dried GO thin film is reduced to RGO by thermal annealing as discussed in Sect. 4.1.3.

4.2 Material Characterization and Discussion

The samples were characterized by SEM (Leo 1530, Carl Zeiss AG, German) to study the morphology of the GO and RGO films. The samples for SEM were prepared by drop-casting GO solution on an RCA clean silicon wafer to form the GO film. The GO film was reduced by a two-step reduction to RGO thin film. The SEM images were taken under an acceleration voltage of 10.00 kV.

Atomic force microscopy (AFM) (DI D3500 Nanoman, Veeco, USA) was used to measure the lateral sizes, thickness, and the numbers of layers of GO and RGO. The samples for AFM were prepared by spin-coating diluted GO dispersion onto an RCA-cleaned silicon substrate. The RGO thin film was produced by reducing GO thin films. The AFM used tapping mode to measure the GO surface and to determine the number of GO layers.

Raman spectroscopy (Ramascope dual-wavelength micro-Raman spectrometer, Renishaw, Canada) was used to further investigate the detailed structure, including lattice and defects changing from GO to RGO. Raman spectra of GO were measured due to the interactions of photons in resonant functionality scattered with the electronic structure. The GO sample was deposited on the Si substrate, and RGO thin film is fabricated by reducing GO.

PL spectroscopy (FLS920, Edinburg Instruments, USA) showed the structure of GO and RGO. The GO solution was drop-casted on the Si substrate and immersed into L-AA solution for 24 h and 48 h to form RGO.

Ultraviolet–visible (UV–Vis) spectroscopy (UV-2501PC, Shimadzu, Japan) was used to measure the transparency of the films.

The electrical conductivity of RGO was measured using a Semiconductor Parameter Analyzer (Agilent 4155C, Agilent Technologies, Canada) connected to a 2-probe station inside a dark, sealed chamber; sheet resistance was measured by the four-point probe (Model RM2, Jandel UK). The semiconductor parameter analyzer measures the resistance of GO and RGO. Then, the conductivity (σ) was calculated using the length and cross-sectional area of the measured portion.

XPS (VG ESCALab 250, Thermo Fisher Scientific, USA) was used to determine the chemical compositions of GO and RGO. The source was monochromated Al K-alpha as the x-ray source. Before the measurement, the plasma-cleaning was applied to remove the contaminants on the surface of GO and RGO films. The charge compensation was applied to the GO sample to reduce the influence of the charge accumulation due to the insulating properties of GO thin films.

Ultraviolet photoelectron spectroscopy (UPS) (VG ESCALab 250, Thermo Fisher Scientific, USA) measured the work function of RGO. The photon source helium discharge lamp emits energy of 21.2 eV in the vacuum UV region. Since UPS only detects several nanometers of the surface, the surface of the sample should be very clean. Therefore, samples were treated by plasma cleaning for 60 s before the measurement. Plasma cleaning can remove hydrocarbon contamination attached on the sample surface due to air exposure.

4.2.1 Visual and Surface Analysis

Figure 4.2 shows the SEM images of the GO and RGO films after different process stages as listed in Table 4.1. Figure 4.2a shows that GO has uneven surfaces and wrinkle layers, indicating defects. As seen in Fig. 4.2b, c, the one-step reduction improved the surface structures of both RGO_{HI} and $RGO_{L\text{-}AA}$ with the RGO_{HI} film exhibiting more uniform and smooth surfaces. This reveals less sheet aggregation and defects after the removal of oxygen functional groups. Figure 4.2d shows that the $RGO_{L\text{-}AA+HI}$ film after the two-step reduction has minimal defects and further improved uniformity compared to $RGO_{L\text{-}AA}$ and RGO_{HI}.

Figure 4.3 shows the layered structure of free-standing GO and RGO film characterized by the cross-section SEM. The thickness of multi-layered RGO film was measured by AFM. In Fig. 4.3a, b, the thickness of A-RGO is about 3.8 μm and S-RGO is 4 μm. The cross-sectional SEM images show the multi-layered structure for both A-RGO and S-RGO, which improves the mechanical strength and flexibility of the RGO thin film. Reprinted with permission from [121]. Copyright 2023, IEEE.

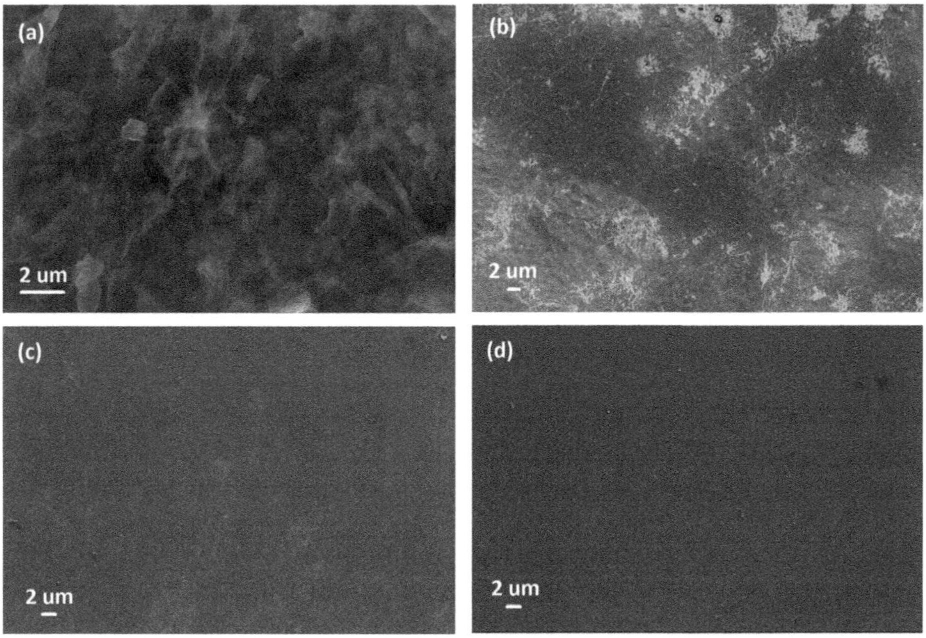

Fig. 4.2 SEM images of **a** GO solution, **b** $RGO_{L\text{-}AA}$ film **c** RGO_{HI} film, **d** $RGO_{L\text{-}AA+HI}$ film

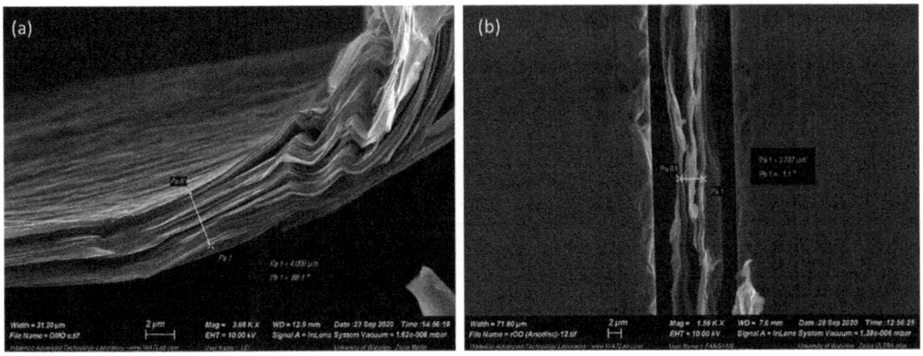

Fig. 4.3 SEM images of the cross-section of the free-standing thin films of **a** S-RGO and **b** A-RGO

4.2.2 AFM Analysis of GO and RGO

Figure 4.4 exhibits AFM images and roughness of the GO and RGO films. AFM in tapping mode was used to determine the thickness and roughness of GO and RGO thin films. Figure 4.4a shows the AFM image of RGO; the height of the RGO film was around 15 nm. Figure 4.4b presents an AFM image of GO; the distance measured by the two points gives the height of GO film, which is around 20 nm. Figure 4.4c shows the roughness of the RGO thin film. Both root mean square (RMS) roughness (Rq) and mean roughness (Ra) are used for detecting the variations in roughness of the peaks and the valleys of the surface to analyze the uniformity of the film. The Rq is 3.1 nm, and the Ra is 1.644 nm. The ratio of roughness to the overall thickness is 8.2%. Figure 4.4d shows the roughness of the GO thin film with Rq at 7.3 nm and Ra at 3.4 nm and the ratio of roughness to thickness is 17%. The reduction of the film thickness from 23 to 20 nm is due to the deoxygenation of the basal plane of GO. It should also be noted that the L-AA reduction, which is the mild conditions, does not contribute to the creation of newer defects. The change in surface hydrophilicity of GO may result from the decreasing oxygen content. The inherent hydrophilicity of GO is due to the hydroxyl, epoxy, and carbonyl functional groups, and their reduction renders the RGO hydrophobic. This is concluded from the decreasing RGO solubility in water and dimethylformamide (DMF) polar solvent.

Figure 4.5 shows the thickness and roughness indicated by the AFM images of free-standing GO and RGO. The root-mean-square roughness (R_q) of A-RGO is about 45.6 nm and S-RGO is about 53.2 nm. The roughness of RGO thin films can be compared by the ratio of R_q to the thickness, which is 0.6% and 1.3% for A-RGO and S-RGO, respectively. Therefore, both RGO films have good uniformity. For A-RGO, forming of the free-standing film is due to the stronger interaction between GO layers than between GO and the substrate [122]. The Anodisc membrane is an Aluminum oxide membrane with

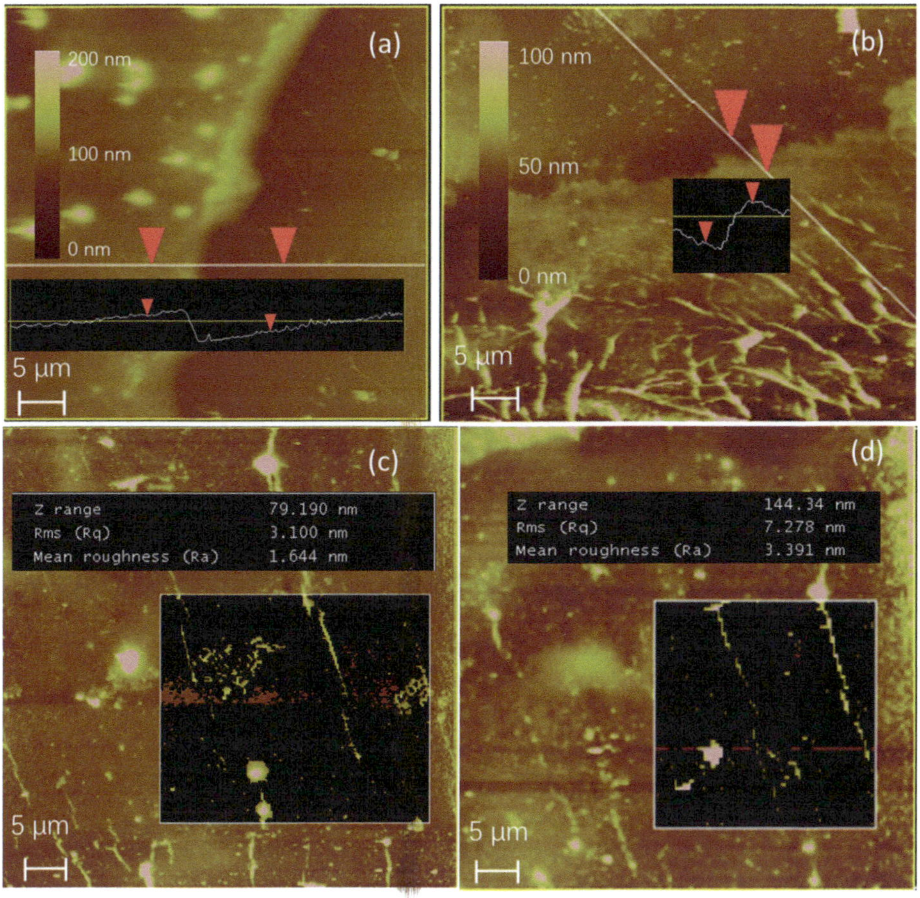

Fig. 4.4 AFM images of **a** RGO; **b** GO thin film, and roughness of **c** RGO and **d** GO

a porous structure, which reduces the contact area of GO and the substrate [123]. The interaction between GO sheets is plane-plane contact geometry, which is stronger than the plane-point interaction between GO and the substrate [122]. Therefore, the drops on the substrate have a small contact angle with the substrate so little pressure difference as in vacuum filtration to cause ripples. For S-RGO, as the water in the suspension evaporates, the concentration of GO in the suspension rises, leading to a significant increase in sheet-to-sheet interactions and finally stacking of sheets on top of one another in the ever-growing deposit, forming a layer-by-layer nanostructure [124]. Therefore, the free-standing film produced by the vacuum filtration method has a smooth surface but is limited in size. Conversely, that fabricated by solution evaporation method has a large area.

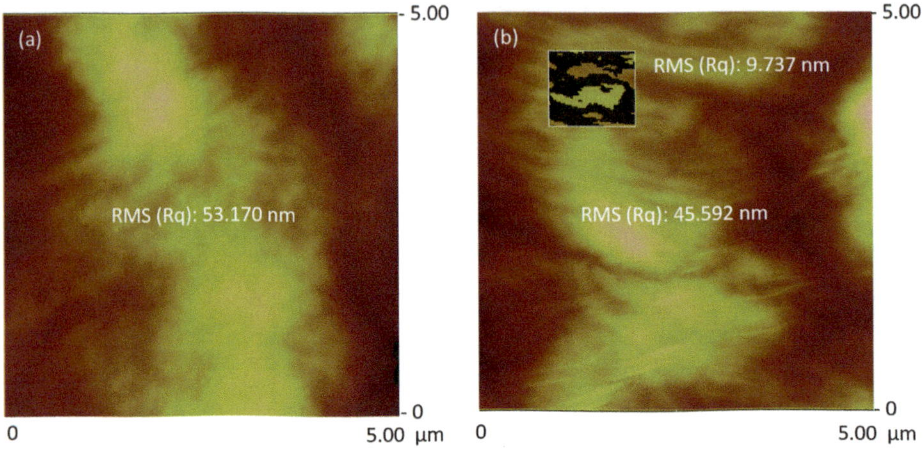

Fig. 4.5 Atomic force microscope (AFM) images of **a** A-RGO and **b** S-RGO

4.2.3 FTIR Analysis of GO and RGO

Figure 4.6 presents the FTIR spectra of the GO and RGO thin films. As shown in the GO spectra, the presence of broadband from 3,000 to 3,700 cm^{-1} indicates the presence of hydroxyl groups, which are due to the stretching vibrational mode of intercalated water. There is a band at around 1730 cm^{-1}, which is assigned to carbon–carbon double bonds. Furthermore, a band at around 1620 cm^{-1} represents carboxyl groups, and a band at around 1220 cm^{-1} is determined to be epoxy groups. The FTIR spectra confirms the presence of oxygenated functional groups in the GO structure. In the RGO spectra, the peaks at around 3000 cm^{-1} from the GO structure significantly decrease or even disappear. This shows that hydroxyl groups can be thoroughly removed by L-AA. However, the epoxy groups at around 1220 cm^{-1} do not change much; the many hydroxyl groups contained in L-AA cannot reduce the epoxy groups. The peak at around 1560 cm^{-1}, corresponding to carboxyl groups, is slightly red-shifted; this is attributed to external influences, such as hydrogen bonds in the water interacting with oxygenated groups to change the bond length [125]. Similarly, the peaks from C=C bonds (around 1720 cm^{-1}) are also slightly red-shifted due to vibration with L-AA. In summary, FTIR results show that L-AA can eliminate a significant amount of hydroxyl groups and some carboxyl groups, but has little effect on epoxy groups, which are weak in GO. In addition, the absence of −OH in the RGO FTIR spectrum indicates that GO has been successfully reduced to RGO.

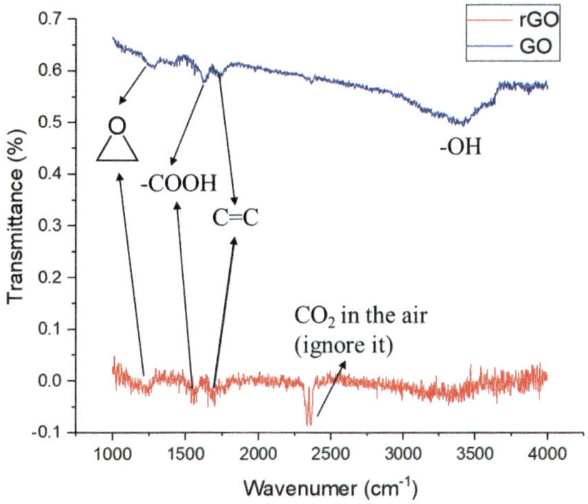

Fig. 4.6 FTIR spectra of GO and RGO

4.2.4 Structural Analysis of GO and RGO

Figure 4.7 shows the Raman spectrum of single-layer graphene, typically containing distinct G and D bands, of the spectra of the GO and RGO films. In general, the G band at 1,580 cm^{-1} or so represents the primary vibrational mode in the single crystal, and the D band at 1,300–1,350 cm^{-1} is related to structural defects and edges appearing in crystallites [107]. It refers to the second elastic scattering by a phonon in the disordered region. In pristine graphene, the D band does not exist because of the crystalline symmetry. In GO, the graphite structure has defects and attached functional groups by the oxidation process resulting in the appearance of a D peak. After reduction of GO, the change of D band shows the disorder and defects restored. Higher intensity of the disorder-induced D-band means degraded crystallinity in graphite-based materials. In addition, a secondary D band, the 2D band, at 2,690 cm^{-1}, located at a double frequency of D band or so is due to two-phonon, double resonance processes and is related to the number of graphene layers due to the electronic band structure splitting in multi-layered graphene [126]. The splitting of in-plane phonon bands is due to the discrete wave vectors. In addition, the 2D band has two peaks $2D_1$ band, the higher energy shoulder and the $2D_2$ band, the lower energy shoulder. This spitting is caused by the π and π^* electrons energy interaction in the next graphitic lattice [126]. For the GO sample, D and G bands are at 1,330 cm^{-1}, and 1,595 cm^{-1}, respectively, indicative of pristine graphite lattice, and the 2D bands are at 2,640 and 2,912 cm^{-1} to show the number of layers and defects. For RGO, the D band is still at 1,330 cm^{-1}, while the G band has a redshift to 1,581 cm^{-1}, which is closer to the theoretical value of 1,580 cm^{-1}. Moreover, the two 2D bands also get closer with one at 2,662 cm^{-1} and another at 2,907 cm^{-1}. Higher intensity of the disorder-induced D-band means degraded crystallinity in graphite-based materials. The ratio between the D and G band intensities, I_D/I_G, is a measure of the graphite-like (sp^2) and diamond-like

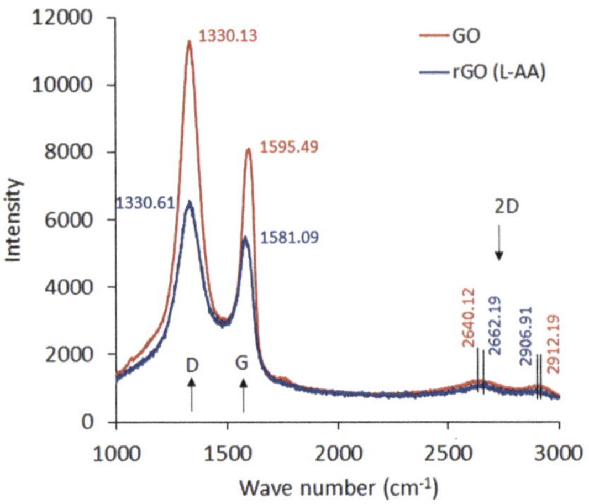

Fig. 4.7 Raman spectra of GO and RGO thin film. Reprinted with permission from [121]. Copyright 2023, IEEE

(sp^3) bond ratios and is indicative of disordered carbon [127]. Figure 4.7 shows that the I_D to I_G ratios for GO and RGO samples are 1.49 and 1.21, respectively. The reduced ratio indicates that the reduction process increases sp^2 and decreases sp^3 bonds.

4.2.5 Chemical Composition of GO and RGO Thin Film

Figure 4.8 shows XPS of GO and RGO thin films. Figure 4.8a, b show the XPS survey spectra of the GO and RGO films, respectively. The peaks around binding energies of 285 eV and 532 eV correspond to the electrons removed by x-rays from C (1 s) and O (1 s) levels, respectively [128]. Each level consists of different functional components. The elemental C/O ratio can be estimated using the C (1 s) and O (1 s) peaks [129]. Data in Fig. 4.8a, b indicate that the C/O ratio was 0.56 for GO and 1.48 for RGO. The increase in the ratio by 1.64 times indicates the effectiveness of the L-AA reduction process. The functional group changes due to the reduction process are further analyzed by considering the C (1 s) signal in both cases. Figure 4.8c, d show the raw XPS data and the deconvoluted peak components (C=C, C–O, C=O) for the GO and RGO samples. In the graphene oxides, the peak centered at 284 eV corresponds to sp^2 carbon and the ones around 286 eV are due to the oxidized carbons [114]. In Fig. 4.8c, the C=C, C–C, and epoxy groups are centered at 283.6 eV, 285.7 eV, and 286.3 eV, respectively. After reduction, the main peak is at 284.5 eV, and small peaks are at 285.5 eV and 288.1 eV (see Fig. 4.8d). Comparing both figures, the peak area in C=C of RGO is larger than oxygenated functional groups, unlike the comparative area of C=C and C–O in GO. Thus, the chemical composition analysis further confirms the effectiveness of L-AA in oxygenated functional group removal and restoration of sp^2 hybridization structure. As for chemical composition changes, XPS results confirm with FTIR results

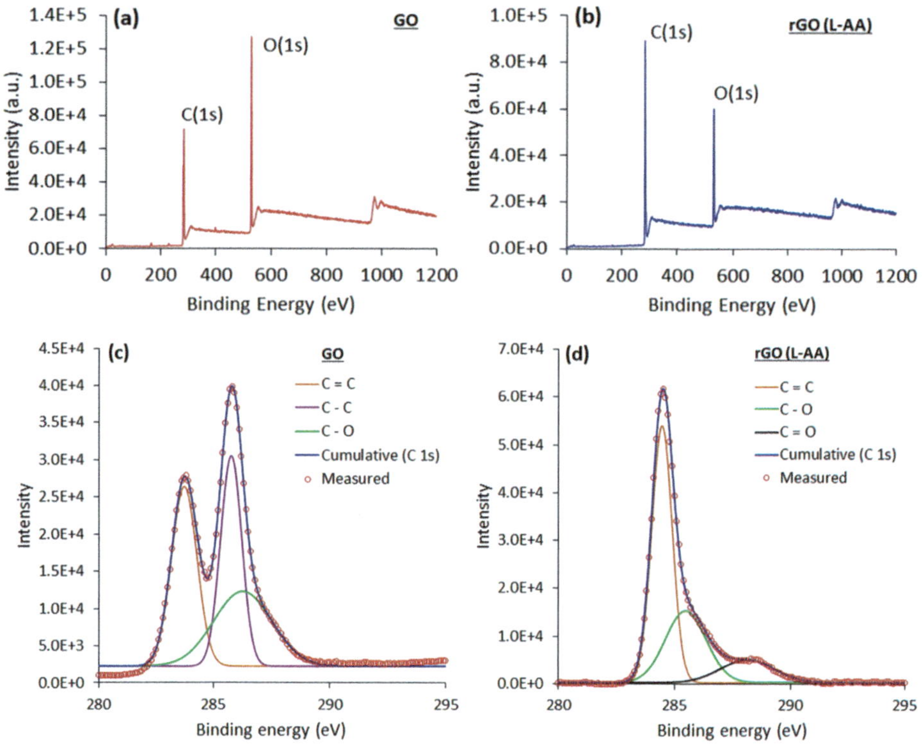

Fig. 4.8 **a** XPS spectra of C 1 s of GO, **b** XPS spectra of C 1 s of $RGO_{L\text{-}AA}$, **c** XPS survey spectra of GO and **d** XPS survey spectra of $RGO_{L\text{-}AA}$. Reprinted with permission from [121]. Copyright 2023, IEEE

that large amount of oxygenated functional groups are removed during the reduction. As for structural change, disappearance of C–C peak in RGO confirms with Raman result that I_D/I_G is decreased. The chemical composition and atomic structure changes bring the smaller sheet distance and larger interaction between layers, as consistent with AFM results. Therefore, delocalized electrons in RGO layers can have a smooth movement, and the product changes from an insulator (GO) to the semiconductor (RGO).

4.2.6 Electrical Conductivity of RGO and Optimization

Electrical conductivity can be regarded as a powerful indicator for deoxygenated graphene. The two-probe IV semiconductor parameter analyzer was used to measure the conductivity. GO is an insulator; hence no value could be detected in conductivity measurements. The significant increase in conductivity confirms that GO (10^{-11} to 10^{-10} S/m)

Table 4.2 Best value of electronic properties of RGO free-standing films

RGO	Sheet resistance (Ω/sq)	Conductivity (S/m)
A-RGO	271.9	967.8
S-RGO	1320	189

was reduced to RGO (>100 S/m). The synthesized RGO with high conductivity of 967.8 S/m can be applied as a TCE in photovoltaics, especially in low-current devices.

Table 4.2 shows the lowest sheet resistance and highest conductivity of A-RGO and S-RGO. The range of the sheet resistance of A-RGO is 381.1 ± 109.2 Ω/sq, and the corresponding conductivity is 752.25 ± 215.55 S/m. For S-RGO, the range of the sheet resistance is 1675 ± 355 Ω/sq, and the corresponding conductivity is 156 ± 33 S/m. The electrical conductivity and sheet resistance depend on the restoration of the carbon sp^2 conjugation system [122]. The electrical conductivity can be influenced by the thickness, layer distance, and so on. Because the GO thin film is formed by natural gravity, a large separation between layers exists as shown in the SEM images (Fig. 4.3). This large distance between each layer allows the barrier for the delocalized electrons to transport. Moreover, during thermal annealing, defects and residual oxygen-containing functional groups influence the restoration of the conjugation system [46]. Rapid heating (>2000 °C/min) thermally reduced GO film achieved higher C/O and electrical conductivity but introduced defects into the graphitic structure during the exfoliation process occurred while the temperature increased [44]. Therefore, the trade-off between the thermal annealing temperature and the structure change may limit the electrical conductivity.

Reduction temperature, reduction time, the concentration of reductants, concentration of precursor, and reductant type are important factors that affect the sheet resistance. In Fig. 4.9, the sheet resistance of $RGO_{L\text{-}AA}$ drops when the temperature increases from 80 to 95 °C. The sheet resistance decreased from 1600 to 111 kΩ/sq when the temperature increased from 80 to 90 °C, but when the temperature increased from 90 to 95 °C, the sheet resistance increased again from 111 to 294 kΩ/sq. The reduction reaction of L-AA and oxygen-containing functional groups is the reversible reaction (Fig. 2.3), heating can promote the reaction to the right side. L-AA is a mild reductant, a high temperature (>70 °C) causes the degradation of L-AA [130].

Figure 4.10 shows the electrical properties of the RGO films, including conductivity, sheet resistance, and optical transmittance to investigate the potential for transparent electrodes. Figure 4.10a shows the electrical conductivity versus reduction time for $RGO_{L\text{-}AA}$ films. The retention time for 16 h resulted in a low conductivity probably because the functional groups cannot be fully reduced. Reduction durations in the range of 18–24 h resulted in a reasonably high electrical conductivity (i.e., 360–480 S/m). For durations of 48 h and 72 h, the conductivity drops again because film hydrophobicity becomes a dominant feature as discussed in the SEM section, and the poor film attachment to the surface results in cracking and flaking in the L-AA solution.

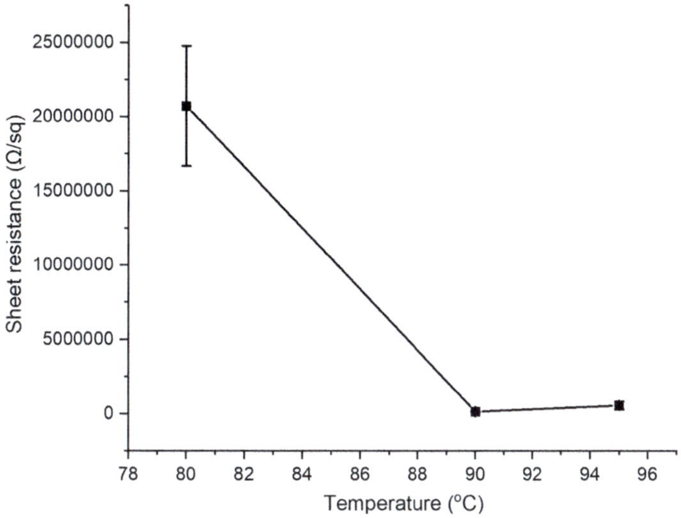

Fig. 4.9 Reduction temperature versus sheet resistance of RGO

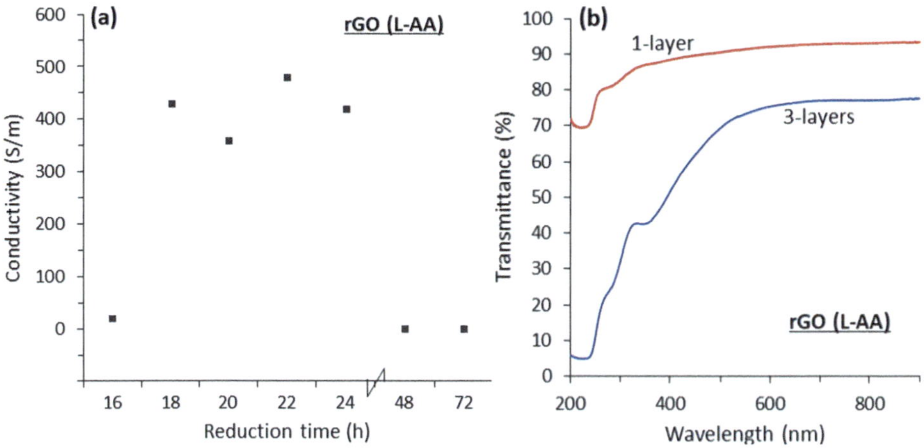

Fig. 4.10 Reduction time vs sheet resistance of RGO

Figure 4.10b shows the optical transmittance of single- and three-layered $RGO_{L\text{-}AA}$ films. The single-layer film with a thickness of 20 nm or so has high transparency over a broad wavelength range, underlining its potential for application in transparent electrodes. Figure 4.10b also shows the transparency from 90% or so for single-layer to 70% or so for the triple-layer film.

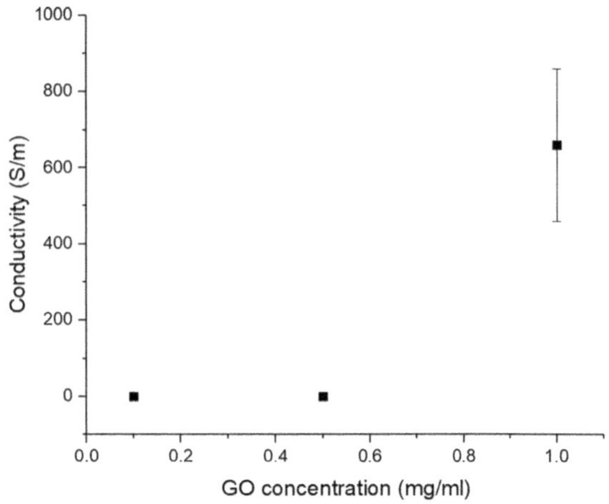

Fig. 4.11 Reduction concentration of GO vs sheet resistance of RGO

The concentration of precursor is also an important factor in the conductivity. Figure 4.11 shows that the conductivity is increasing with an increasing concentration of GO. The concentration of 0.1 mg/ml or 0.5 mg/ml of GO can produce a thin and non-uniform film (Fig. 4.2). After reduction, lattice defects were produced so that delocalized electrons travelling paths were blocked by shortening the ballistic transport path and creating scattering centers, which invariably impact the product's electrical properties [131]. For a high concentration of 1 mg/ml, a uniform film can be produced (Fig. 4.2). The proper GO concentration can produce films with few defects, which lengthen the delocalized electron's path, so the conductivity can be improved.

Figure 4.12 shows that the reductant used also plays an important role in sheet resistance. Different reductants were tested to further explore ways, including the development of a two-step reduction process with L-AA and HI (both liquid and vapor) reductants, to improve the electrical properties. Figure 4.12 shows the measured sheet resistance of RGO films (all single-layer, comparable thickness of 20 nm or so) produced using various reductants. The process with L-AA in the first step and HI in the second step results in the lowest sheet resistance 59.8 kΩ/sq. From the reduction reaction (Figs. 2.3 and 2.4) L-AA and HI can both remove hydroxy and epoxy groups, and their combination seems to be the most efficient to reduce GO.

4.2.7 Work Function of RGO

4.2.7.1 Introduction of UV Photoelectron Spectroscopy (UPS)

Besides conductivity, the work function is an important property of RGO, which can be measured by UPS. The work function is defined as the minimum energy to extract electrons from the materials. The work function is not an intrinsic property because it is affected by the surface impurities, film uniformity, crystal orientation, and so on [132].

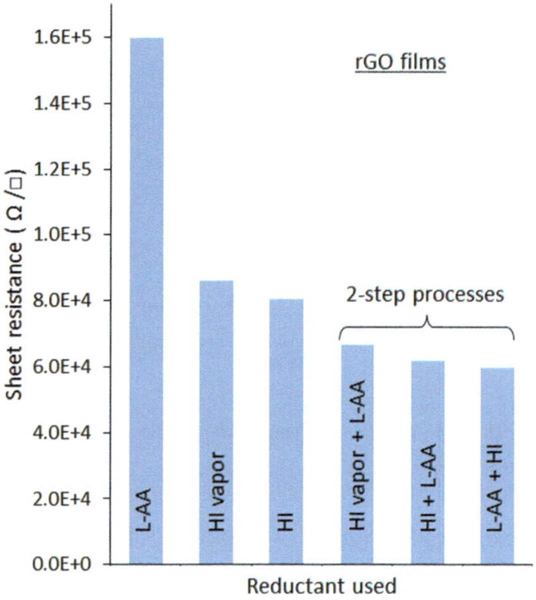

Fig. 4.12 The sheet resistance of RGO thin-film is reduced by different reductants. Reprinted with permission from [121]. Copyright 2023, IEEE

To measure the work function, UPS can measure kinetic energy (KE) or binding energy (BE), and the work function can be calculated using Eq. 4.1 [132],

$$\phi = h\nu - \left(E_F - E_{cutoff}\right) \tag{4.1}$$

where $h\nu$ is the photon energy of the excitation light, E_F is the Fermi level energy, and E_{cutoff} is the inelastic high binding energy cut-off energy.

A 500 nm e-beam evaporated Alumium film is measured as the reference sample to calculate the work function from UPS spectra. Figure 4.13 shows the UPS spectrum of Al. In Fig. 4.13b, the secondary cut-off energy is at 19.5 eV, so the work function is calculated as 21.2 − (19.5 − 2.5) = 4.2 eV, which is the same as the theoretical value [133].

4.2.7.2 Work Function of GO and RGO

Figure 4.14 shows the UPS spectra of the $RGO_{L\text{-}AA}$ and $RGO_{L\text{-}AA + HI}$ films. hν was 21.22 eV because helium I discharge was used as the UV source. A voltage bias of 5 eV was applied to the measurements with RGO films. In Fig. 4.14b, the black arrows show $E_{MAX} - E_{MIN}$ and E_{VB} of the spectrum of $RGO_{L\text{-}AA + HI}$. The cut-off edge is calculated according to the procedure described in [134], by focussing on the circled region of Fig. 4.14a and finding the point corresponding to the largest slope by taking the first derivative. Figure 4.14b shows that the cut-off values are 10.08 eV and 8.64 eV for the $RGO_{L\text{-}AA}$ and $RGO_{L\text{-}AA + HI}$ films, respectively. The Fermi edge is located at 21.22 + bias, i.e., 26.22 eV. The work function of $RGO_{L\text{-}AA}$ is thus calculated as 5.1 eV

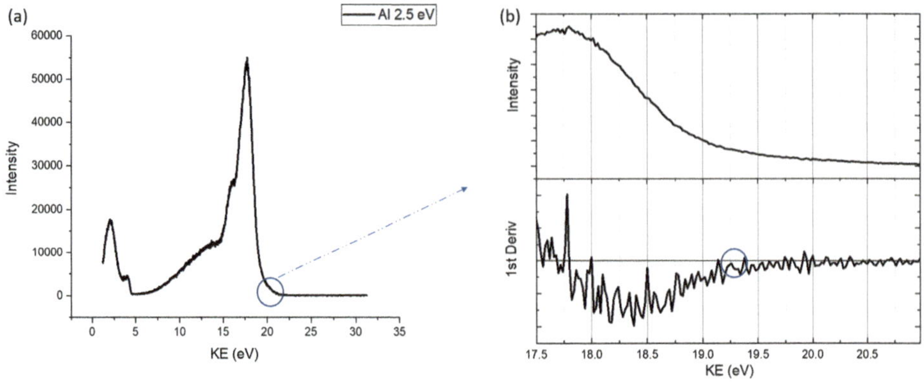

Fig. 4.13 **a** UPS spectrum of Al by 2.5 eV bias; **c** first derivative of the intensity of Al from 17.5 to 21.5 eV

[i.e., 21.22 – (26.22 – 10.08) = 5.1 eV], and that for $RGO_{L\text{-}AA + HI}$ is 3.64 eV [i.e., 21.22 – (26.22 – 8.64) = 3.64 eV]. The progressive removal of oxygen functional groups (carboxyl, epoxy, and hydroxyl), as also seen in XPS, leads to the reduction in work function (ϕ). The reduction in ϕ by removal of O-functional groups in GO via prolonged UV irradiation was also reported by others [135]. UPS on the un-reduced GO films was performed, for which a bias of 10 eV had to be applied, which resulted in a broader and slightly low-intensity UPS spectrum. As expected, a higher ϕ value ($\approx$5.9 eV) was obtained for GO following a similar procedure.

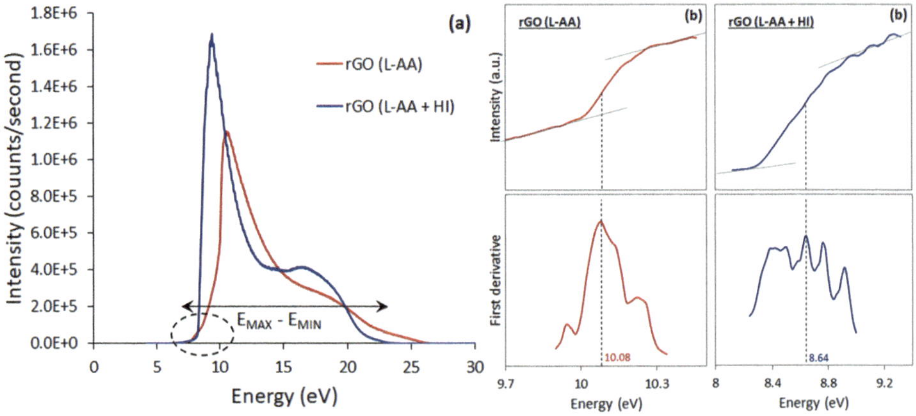

Fig. 4.14 **a** UPS spectra of the $RGO_{L\text{-}AA}$ and $RGO_{L\text{-}AA + HI}$ films measured with helium I discharge and 5 eV bias. **b** Determination of the energy cut-off focussing on the circled region of **a**, following the procedure described in [136]. Reprinted with permission from [121]. Copyright 2023, IEEE

Overall, the drop in ϕ from GO to $RGO_{L\text{-}AA}$ to $RGO_{L\text{-}AA+HI}$ shows that a broader range of work function tunability in RGO can be achieved by controlling the reduction processes. This can lead to enhanced applications, including carrier-selective electrodes. Finally, the UPS method is particularly sensitive to surface contamination, but the precaution was taken in all experiments by plasma-cleaning the samples within 60 s before the UPS measurements. The procedure adopted to estimate ϕ from measured UPS data may also add minor uncertainty. Nevertheless, the relative range of tunability by employing the two-step reduction process is established. The work function depends on the presence of different oxygenated functional groups. The work function is dropped from GO to $RGO_{L\text{-}AA}$ to $RGO_{L\text{-}AA+HI}$ resulting from the presence of surface dipole moments forming from the oxygen functional groups which disrupt the π-conjugation network. The concentration of functional groups is one of the factors. In DFT calculation, the concentration of carbonyl groups increases from 1.5 to 20%, the WF of GO changes from 4.4 to 6.8 eV [78].

4.2.8 Photoluminescence Analysis of RGO

GO and RGO can both be considered disordered forms of the base hexagonal structure of graphene, resulting in non-zero bandgaps and desirable photoexcitation properties [137]. The GO materials, consisting of variable sp^2 and sp^3 fractions, can exhibit photoluminescence (PL) resulting from localized electron–hole pair recombination in the sp^2 clusters [138]. Figure 4.15a, b show the photoluminescence (PL) spectra of the GO and RGO samples, respectively. Both figures show the measured data and the deconvoluted fitting peaks. The PL of GO has centered at 520 and 760 nm in Fig. 4.15a. To prepare the RGO for PL measurements, GO solution (4 ml) was mixed with L-AA (2 ml) and DMF solvent (4 ml) and the reduction process was carried out. Figure 4.15b shows the PL spectra of $RGO_{L\text{-}AA}$, with the main peak position considerably blue-shifts to 430 and 465 nm, and the small peak shifting to 600 nm. The main peak also becomes narrower. The reduction processes in GO produce new sp^2 clusters and decrease the number of disorder-induced defect states in the gap [139]. The former explains the blue-shifting and the latter the narrowing of the main peak. This also agrees with what was observed from the Raman analyses. Overall, the reduction process offers tunability with varying sp^2 and sp^3 fractions and restores the structure with reduced defects in RGO.

4.3 Summary

In summary, a mild two-step method is demonstrated to produce RGO. The two-step reduction process reduces GO to synthesize solution dispersed RGO, which was then formed into films with a thickness of 20 nm or so. The GO starting material was

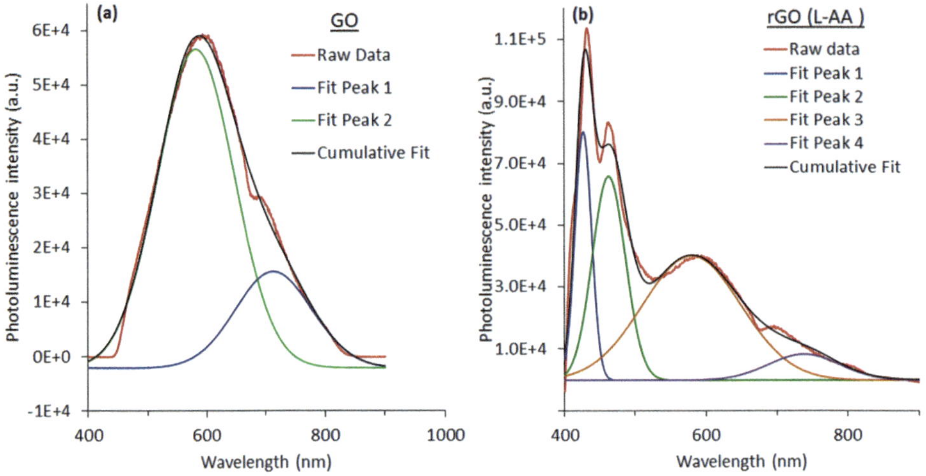

Fig. 4.15 Photoluminescence spectra of the **a** GO and **b** $RGO_{L\text{-}AA}$ samples show the measured data as well as the deconvoluted fitting peaks. Reprinted with permission from [121]. Copyright 2023, IEEE

formed from inexpensive commercially available graphite flakes. The uniform RGO films produced from a defective GO were characterized by structural, optical, and electrical properties. The combination of L-ascorbic and hydroiodic acids help eliminate residual-oxygenated functional groups after the first reduction. The film transparency was 90% or higher over a wavelength of 450–800 nm. The work function of the RGO material, characterized by ultraviolet photoelectron spectroscopy, is in the range of 3.6–5.1 eV. Thus, the choice of reduction process can effectively control the work function of RGO.

5 Device Application of Reduced Graphene Oxide

5.1 RGO in MOS Devices: Design and Functionality

The MOS device was fabricated to evaluate C-V characteristics with RGO as the top contact. The MOS device is a sandwich structure with a thin layer of oxide in the middle of gate metal and Si substrate. In the reference device, Al is used as the gate metal and back contact, 10 nm SiO_2 was deposited on the p-Si substrate, which is to compare with RGO-based MOS devices. In this device, fabrication and operation of MOS device built using RGO as the gate metal. RGO exhibits high conductivity and work function which is a better choice for the gate metal instead of Al. The area of RGO is controlled to be as small as possible to transport most carriers (Fig. 5.1).

5.2 Fabrication Process of MOS Devices with RGO

5.2.1 MOSCAP Device Fabrication Procedure

The device fabrication process is as follows. A p-type silicon wafer with (100) surface orientation doped by boron with 1–5 Ω cm resistivity and 525 ± 25 μm thickness was first passed through RCA 1, hydrofluoric acid (HF) etching, RCA 2 and HF etching process in time sequence. Then, before the wafer went to the Rapid Thermal Processor (RTP600S, Modular Process Technology Corp., USA) system for thermal growing silicon dioxide (SiO_2), the wafer was dipped in 2% HF solution for 1 min to remove the naturally grown oxide layer. SiO_2 was then grown with a 60 s ramp to 1000 °C in N_2 ambient, then a 5 s hold at 1000 °C in both N_2 and O_2, 25 s hold at 1000 °C with pure O_2, and 5 s hold at 1000 °C in both N_2 and O_2. The sample was then cooled down to room temperature in N_2 ambient in 60 s. The thickness of the SiO_2 layer measured by the ellipsometer (M-2000,

F. Sun, *Synthesis and Device Applications of Graphene Derivatives and Quantum Dots*,
Synthesis Lectures on Green Energy and Technology,
https://doi.org/10.1007/978-3-031-56403-1_5

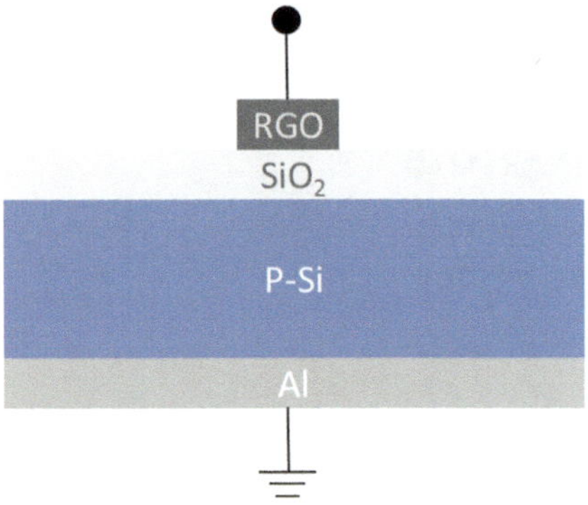

Fig. 5.1 Device schematic of RGO-based MOSCAP device

J. A. Woollam, USA) was 10 nm. Then, before depositing back contact on the backside of the wafer, the oxide on the backside of the wafers was removed in 2% HF solution for 1 min using a PTFE holder that seals one side of the wafer using the O-rings. A 320 nm thick aluminum layer was thermally deposited on the backside by a thermal evaporator (NCII-6EB-T, Intlvac Inc, Canada) as the back contact. The deposition rate is about 2 Å/s at the power of 12.8%. Finally, the back contact was sintered using an ultraclean furnace (Tytan, Tylan Corporation, USA) in forming gas (N_2/H_2) at 400 °C for 30 min.

5.2.2 Fabrication Process of RGO as Gate Metal

The RGO gate metal was fabricated by drop-casting GO on Si substrate and reducing it to RGO. Then, 2 μl 0.78 mg/ml GO was drop-casted on the front side of the wafer as the top contact through a Kapton tape mask to control the area to 0.007 cm^2. Finally, the film was sintered to 300°C using a hot plate (Cimarec+™ Stirring Hotplates Series, Thermo Fisher Scientific, USA) for 3 h.

5.2.3 Fabrication Process of Al as Gate Metal

The Al gate metal was fabricated by thermal annealing. The area of Al is the same as RGO at 0.007 cm^2. Thickness at 110 nm Al was deposited at the deposition rate of 2 Å/s by setting the power at 12.8% for 30 min. Then, the top contact was sintered at 400 °C for 30 min with forming gases (N_2/H_2).

5.2.4 Characterization Techniques

Semiconductor Characterization System (Keithley 4200A-SCS, USA, equipped with the 4210-CVU Integrated C-V Option software) was used to make measurements of C-V. The accumulation, depletion, and inversion regions can be identified from the C-V curve. The C-V curve was measured at a frequency of 1 MHz with one probe on the RGO gate metal and another probe on the stage.

Raman spectroscopy (Renishaw micro-Raman spectrometer, Canada) was used to explore the defects and structure from GO to RGO. Excitation was accomplished by using laser wavelengths at 633 nm (red, He–Ne). Raman spectroscopy was equipped with a reflected light microscope which enabled the analyzed area of 5 μm.

XPS (Thermo Fisher Scientific, VG ESCALab 250, USA) showed the oxygen-containing functional groups and the carbon to oxygen (C/O) ratio of RGO. The surface of the sample was first sputtered clean for 60 s to remove the contaminants. The source was a monochromated Al K-alpha laser.

UPS (Thermo Fisher Scientific, VG ESCALab 250, USA) measured the work function of RGO. The photon source was using a helium discharge lamp which emits energy of 21.2 eV in the vacuum UV region. The samples were plasma-cleaned for 60 s to remove the contaminants on the surface before measurements.

Four-point probe (Jandel RM2, UK) measured the sheet resistance of RGO film deposited on the SiO_2/Si substrate. Two probes in the middle measured the current and the other two probes measured the voltage that was applied to the samples. The spacing between the probes was about 1 mm. The sheet resistance was measured by Eq. 5.1

$$R_s = \frac{kV}{I} \tag{5.1}$$

where k is the geometric factor, for the semi-infinite thin sheet, k = π/ln2 = 4.53 [140].

5.3 Device Characterization and Discussion

5.3.1 Capacitance–Voltage (CV) Measurement of MOSCAP

In Fig. 5.2, the C-V curve characterizes MOSCAP with Al and RGO as gate metal, respectively. The theoretical oxide capacitance is calculated as in Eq. 5.2

$$C_{ox} = \frac{A\varepsilon_{ox}}{t_{ox}} = \frac{3.9 \times 8.85 \times 10^{-14} \times 0.007}{100 \times 10^{-8}} = 2.5 \times 10^{-9}\,\mathrm{F} \tag{5.2}$$

where A is the area of the oxide layer, ε_{ox} is the dielectric constant of SiO_2 and t_{ox} is the thickness of the oxide layer. The oxide capacitance measured by the C-V curve is about 5E-9 F as shown in Fig. 5.2a. This result shows good agreement with the theoretical

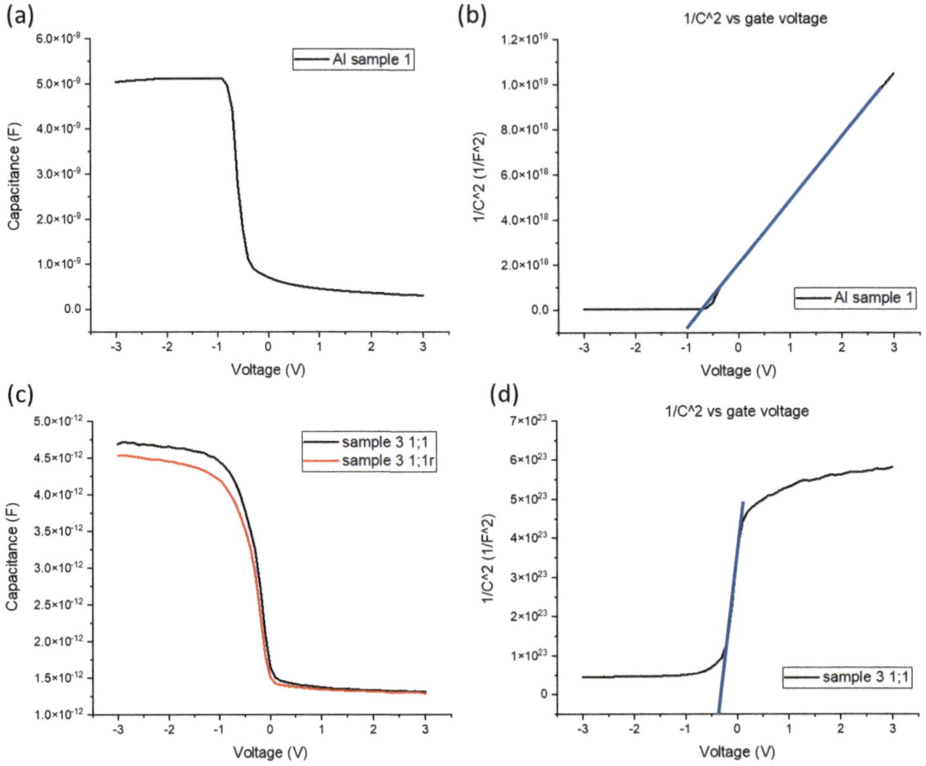

Fig. 5.2 C-V curve of MOSCAP with **a** Al and **b** RGO as gate metal and flat band voltage of **c** Al and **d** RGO

value in the same order of magnitude. Moreover, the accumulation, depletion and inversion regions are shown in the C-V curve. The flat band voltage (V_{FB}) is calculated from the slope of 1/C^2 verse gate voltage, which is −1 eV [141]. In Fig. 5.2b, the MOSCAP device with RGO gate metal shows the accumulation, depletion, and inversion region. The flat band voltage is calculated as −0.4 eV. The oxide capacitance has a small disagreement with the calculated value due to the contact resistance between RGO and the oxide layer. The conductivity of RGO cannot be comparable with Al, so the measured oxide capacitance is not that accurate. Although there is some mismatch, the shape of the C-V curve is consistent with the high-frequency C-V characteristic.

5.3.2 Structural Analysis of RGO

Figure 5.3 shows the Raman spectra of RGO gate metal compared with GO to show the structure change from reduction. The D peak of GO is centered at 1332.84 cm^{-1}, and

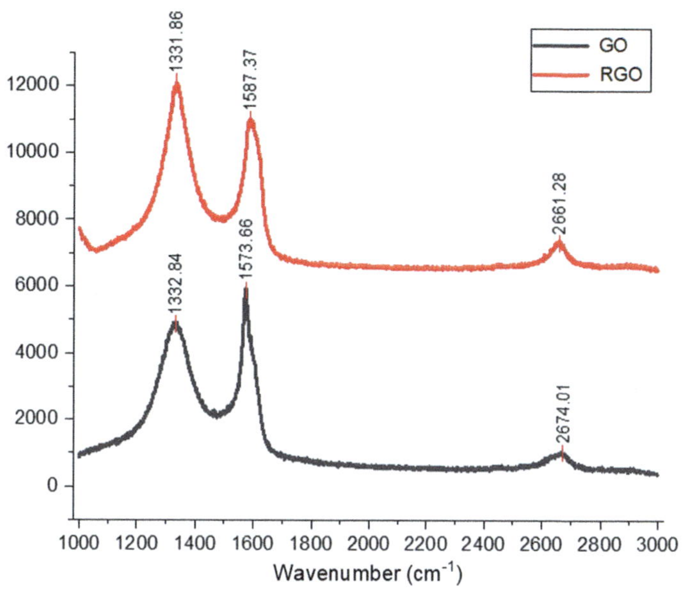

Fig. 5.3 Raman spectra of RGO gate metal

the G peak is at 1573.66 cm^{-1}, while for RGO, the D peak is centered at 1331.86 cm^{-1}, and the G peak has a blue shift to 1587.37 cm^{-1}. The blue shift is due to the frequency of phonons interacting with the incident photon increased due to hole doping, which is usually in the thermal annealing process [142]. The 2D peak is shifted from 2674.01 to 2661.28 cm^{-1}. Moreover, the intensity ratio of the D and G bands (I_D/I_G) can characterize the degree of disorder in graphene. I_D/I_G is non-monotonic as the function of the average distance between defects (L_D). In small L_D at around 4 nm, I_D/I_G increases with the L_D increasing, but in large $L_D > 4$ nm, I_D/I_G decreases with increasing L_D [44]. Therefore, the increasing ratio is caused by the small size of sp^2 domains in the GO and decreasing ratio is caused by the large size. The I_D/I_G of GO and RGO are 0.8 and 1.23, which shows the increase in the small size domains. Thermal annealing is different from chemical reduction, it also removes carbon atoms and splits the graphene sheets into small pieces [131].

5.3.3 Chemical Compositions of RGO

Figure 5.4 illustrates the XPS spectra of RGO gate metal in C 1 s and the survey spectrum to show the chemical composition of RGO. In Fig. 5.4a, the deconvoluted C 1 s spectrum shows the functional groups in RGO. The sp^2 hybridized carbon is centered at 284.57 eV, C-O is at 285.73 eV, and π-π^* satellite is at 291.33 eV. The satellite structure is due

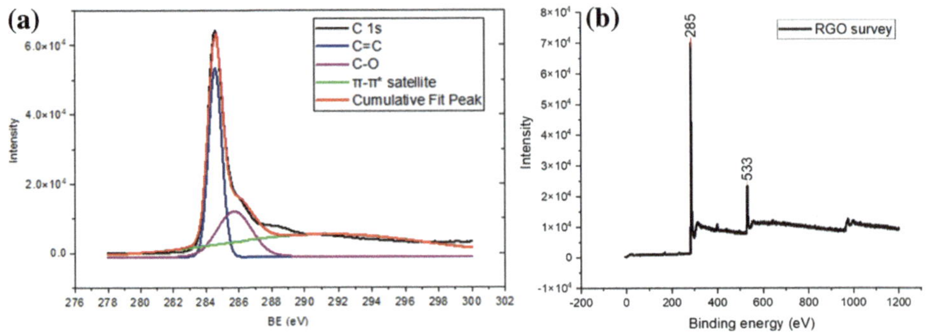

Fig. 5.4 XPS spectra of RGO gate metal **a** C 1 s and **b** survey spectrum

to the extended delocalized electrons (aromatic rings) in samples, usually 6 eV from the main peak of C 1 s [143]. The result indicates the decomposition of oxygen-containing functional groups and aromatic carbon structure restored in RGO. In Fig. 5.4b, the survey spectrum shows the C 1 s peak at 285 eV and the O 1 s at 533 eV. The carbon to oxygen ratio (C/O) is calculated as 3, which is higher than that of chemically RGO. The high C/O ratio indicates a high electrical conductivity.

5.3.4 Electrical Conductivity of RGO

Figure 5.5 presents the sheet resistance and electrical conductivity of RGO gate metal at different concentrations of RGO. The electrical conductivity of GO at 1.04 mg/ml and 0.78 mg/ml is at around 4000 S/m (250 Ω/sq) and 2333 S/m (558.5 Ω/sq), respectively. The high electrical conductivity makes RGO the gate metal comparable with Al. For GO at 1.04 mg/ml, the sheet resistance is measured from the edge to the center, then to the edge. The electrical conductivity is highest in the center of RGO and decreases to the edge. The drop-casting film of RGO has a higher thickness in the center than at the edge. Compared to chemically reduced RGO, the sheet resistance and electrical conductivity increased significantly. This result also confirms higher C/O for thermally reduced RGO. However, the thermal annealing process decreases the attachment between the film and substrate. The large-sized film covered the substrate is easily peeled off; therefore, thermal annealing limits the area of the film.

5.3.5 Work Function of RGO

Figure 5.6 measures the work function of RGO gate metal by UPS spectrum. Figure 5.6a shows the UPS spectrum of RGO and Fig. 5.6b shows the first derivative of the UPS

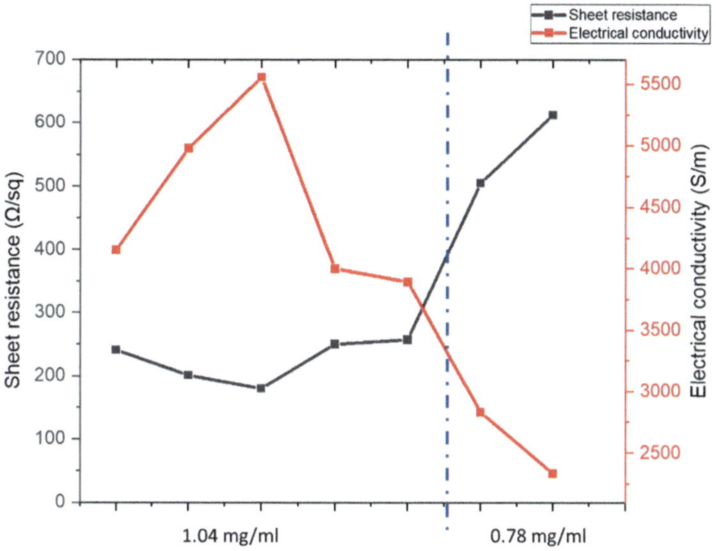

Fig. 5.5 Sheet resistance of RGO gate metal synthesized by different concentrations of GO

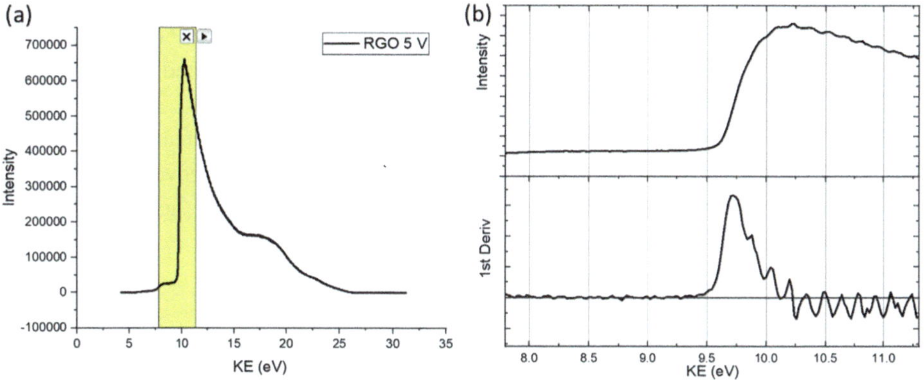

Fig. 5.6 **a** UPS spectrum of RGO gate metal measured with He I discharge and 5 eV bias and **b** Determination of cut-off point focusing on the yellow region

spectrum to find the cut-off point. The cut-off point is at 9.7 eV and the work function of RGO is calculated as:

$$21.2 - (21.2 + 5 - 9.7) = 4.7\,\text{eV} \quad (5.3)$$

In the MOS device, the work function difference between semiconductor and metal indicates the contact resistance. The work function difference is calculated as:

$$W_{MS} = W_M - \left(W_s + \frac{E_{BG}}{2} - \varphi_B\right) = W_M - \left(W_s + \frac{E_{BG}}{2} - \frac{KT}{q}\ln\left(\frac{N_{bulk}}{N_i}\right)\right) \tag{5.4}$$

where W_M is the work function of gate metal, W_S is the substrate material work function, E_{BG} is the substrate bandgap, and Φ_B is the bulk potential and N_{bulk} is the bulk doping concentration. For Al,

$$WMS = 4.1 - 4.15 + \frac{1.12}{2} - 1.46 = -0.95\,\text{eV} \tag{5.5}$$

and for RGO,

$$WMS = 4.7 - 4.15 + \frac{1.12}{2} - 1.46 = -0.35\,\text{eV} \tag{5.6}$$

The work function difference is smaller than Al. Therefore, the high work function of graphene can match better than the traditional metal.

5.4 Summary

A simple MOSCAP device with RGO as gate metal was fabricated based on the optimized process for RGO, SiO_2 layer and Al deposition. The C-V measurement clearly showed accumulation, depletion, and inversion region of the MOSCAP device. The RGO structure and chemical composition were characterized by Raman and XPS. Compared with chemically reduced RGO, thermal annealing can remove most oxygenated functional groups and have higher C/O. The electrical conductivity is enhanced to 5000 S/m due to the recovered structure and functional groups removal. Moreover, the high work function of RGO shows a low work function difference between the gate metal and the semiconductor.

6 Synthesis of Graphene Quantum Dots for Photovoltaics

6.1 Process Development for Top-Down Synthesized GQD

In the oxidative cleavage process, GQD was first oxidized by concentrated H_2SO_4 and HNO_3. 300 mg graphite was added in a 100 ml beaker on a hot plate with stirring. Then, 60 ml H_2SO_4 and 20 ml HNO_3 were slowly added. The mixture was stirred for approximately 20 min until the temperature cooled to room temperature. Then, the temperature was set to 125 °C, 150 °C, 175 °C and 200 °C for 25 h for each sample. Then, 40 ml product was cooled down to room temperature purified by centrifuging at 6000 rpm for 5 min and filtrated by a 0.02 μm filter. The prepared samples were called GQD-125 °C-1, GQD-150 °C-1, GQD-175 °C-1 and GQD-200 °C-1.

Then, the product was further oxidized by hydrogen peroxide. Table 6.1 shows the two-step preparation temperature and time. 5 ml 30% H_2O_2 was added to the remaining 40 ml solution. Then, the solution was separated into two parts. For One part, the solution was mixing solution in 10–30 min. These samples were called GQD-125 °C-2, GQD-150 °C-2, GQD-175 °C-2 and GQD-200 °C-2. For the other one, the solution was heated to 100 °C for 24 h. These samples were called GQD-125 °C-3, GQD-150 °C-3, GQD-175 °C-3 and GQD-200 °C-3. The two kinds of products were purified by centrifuging at 9000 rpm for 5 min and filtrated by a 0.02 μm filter. Purified GQDs were redispersed into the water with sodium hydroxide (NaOH) until pH at 7. Then, the solution was dialyzed using a dialysis tube (Sigma-Aldrich, molecular weight cut-off (MWCO) 1 kDa, PURG10015) to obtain GQDs overnight. Finally, the prepared samples were stored at 4 °C in a fridge. Table 6.1 shows the GQD preparation process and sample names.

F. Sun, *Synthesis and Device Applications of Graphene Derivatives and Quantum Dots*,
Synthesis Lectures on Green Energy and Technology,
https://doi.org/10.1007/978-3-031-56403-1_6

Table 6.1 Two-step GQD preparation

Name	Strong acid	Strong acid temperature (°C)	H_2O_2	H_2O_2 temperature (°C)	H_2O_2 reaction time
GQD-1	H_2SO_4/HNO_3	125, 150, 175, 200	No	–	–
GQD-2	H_2SO_4/HNO_3	125, 150, 175, 200	Yes	25	30 min
GQD-3	H_2SO_4/HNO_3	125, 150, 175, 200	Yes	100	30 min

6.1.1 Materials Characterization Techniques

Raman spectroscopy (Renishaw ramascope dual-wavelength micro-Raman spectrometer equipped with a reflected light microscope, Canada) with a Raman shift resolution of 0.5 wavenumbers was used to investigate the structure and defect in crystallinity of GQDs; laser wavelengths were used for 633 nm as the excitation wavelength (red, He–Ne).

XPS (Thermo Fisher Scientific, VG ESCALab 250, USA) is used to determine the functional groups of the sample surface. The sample is prepared by drop-casting GQD solution on the ITO-coated glass. Before measuring the sample, the surface of the sample is plasma cleaned by sputtering for 60 s.

Photoluminescence (PL) spectroscopy (Edinburg Instruments, FLS920, USA) is used to measure PL emission and excitation spectra of GQDs [90]. The samples are in the form of solution and thin film. The excitation wavelength is 360 nm with 8 nm bandwidth in the setting for both excitation and emission.

Time-resolved spectroscopy is used to measure the fluorescence lifetime of GQDs. It includes the fluorescence spectrometer (Edinburgh Instruments, FLS920, USA) with a pulsed LED (Edinburgh Instruments, EPLED-380 USA, 375 nm, 5 M Hz, 943.3 ps pulse width). The fluorescence decay curve is analyzed by the time-correlated single-photon counting (TCSPC) method. Single-photon events are detected in theory, and their arrival time is connected to the laser pulse that excited the sample. This procedure may be repeated several times using a pulsed laser with a high repetition rate, resulting in a photon dispersion over time and space [144]. The decay data were fitted by the fluorescence analysis software (FAST Version 3.0. Edinburgh Instruments Ltd.). GQD film was prepared by drop-casting the neutralized solution on the ITO-coated glass. GQD solution is measured before post-treatment.

UV–Vis spectroscopy (Shimadzu, UV-2501PC, Japanese) is used to measure the absorption of the sample combined with PL to determine the size. The optical transmittance of the films was measured in the wavelength range of 200–900 nm. The GQD solution samples were prepared in a cuvette.

6.2 Materials Characterization Results and Discussion

6.2.1 Photoluminescence of GQDs

Figure 6.1 presents the PLE spectra of GQD-1, GQD-2 and GQD-3 at 125 °C, 150 °C, 175 °C and 200 °C to show the PL enhancement by temperature and further oxidation. In Fig. 6.1a, all PL peaks of GQD-1 have the same spectral line shapes and peak position centered at around 504 nm. The PL intensity of GQD at 150 °C is the highest, then GQD at 175 °C and 125 °C are similar, but the lowest one is GQD at 200 °C. From the result, 150 °C is the optimized temperature to prepare GQDs only using concentrated sulfuric acid (H_2SO_4) and nitric acid (HNO_3). However, there is not a significant development in PL intensity for only modulating the temperature. Therefore, modification in the oxidative sources is required. In Fig. 6.1b, GQDs were further oxidized by slowly adding H_2O_2 solution into the reaction system, named GQD-2. The PL peaks center of GQD-150 °C, GQD-175 °C and GQD-200 °C have a blue shift to 490 nm, except for GQD-125 °C, the PL center of which is at 500 nm. The intensity of GQD-175 °C is the highest, which is 1/3 higher than GQD-150 °C and 200 °C, and 3 times higher than GQD-125 °C. Therefore, the result shows a significant increase in PL intensity of GQD-1 prepared at different temperatures after adding H_2O_2 except for GQD-125 °C. Compared to GQD-1, the PL intensity of GQD-150 °C-2, GQD-175 °C-2 and GQD-200 °C-2 increases significantly at 2.5, 4 and 3 times, respectively. In Fig. 6.1c, the peaks of GQD-3 are centered at around 490 nm. The PL intensity of GQD-175 °C-3 is the highest at around 4E6, the rest of GQD-125 °C, GQD-150 °C and GQD-200 °C are at around 3.8E6, 3.6E6 and 3.5E6, respectively. The PL intensity difference among GQD-3 at different temperatures is smaller than that in GQD-2. This can be explained by adding H_2O_2 to the reaction; the oxidation is stronger to cleave GQD into smaller pieces. Therefore, the multiple factors affect the PLE spectra, and the further chemical composition is discussed in Sect. 6.3.2.

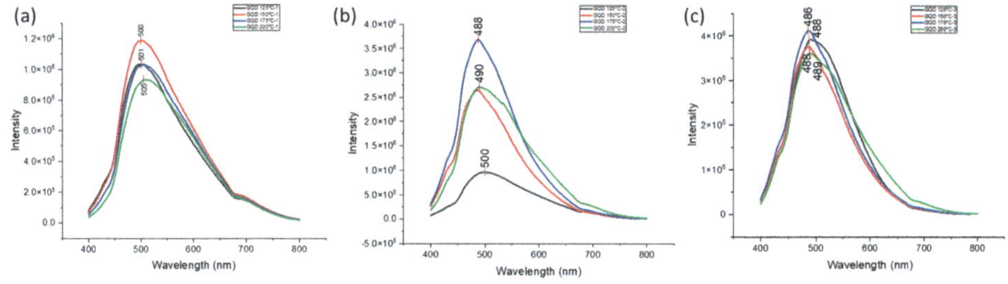

Fig. 6.1 PL spectra of **a** GQD-1, **b** GQD-2 and **c** GQD-3 at 125 °C, 150 °C, 175 °C and 200 °C

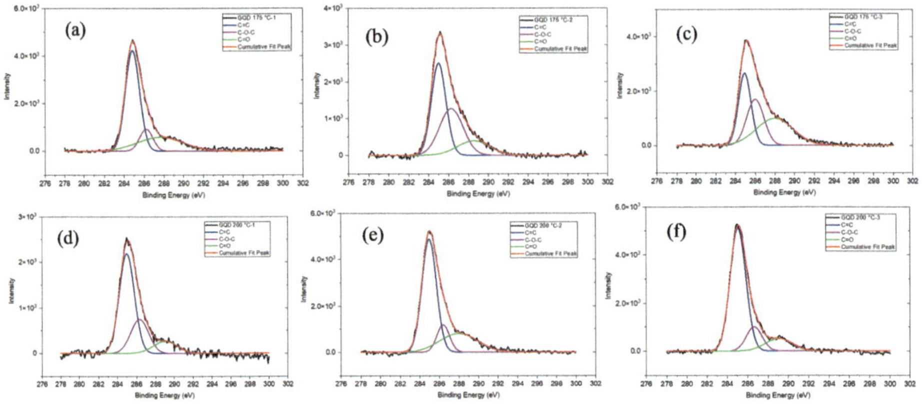

Fig. 6.2 XPS spectra of GQD. Deconvolution of the C 1 s core-level spectra for **a** GQD 175 °C-1, **b** GQD 175 °C-2, **c** GQD 175 °C-3, **d** GQD 200 °C-1, **e** GQD 200 °C-2 and **f** GQD 200 °C-3

6.2.2 Chemical Composition of GQDs

The chemical composition of GQD can be characterized by XPS shown in Fig. 6.2. The measured deconvoluted C 1 s spectra have 3 components, corresponding to C=C/C–C at a binding energy of 284.8 eV, C–O–C at around 286 eV, and C=O at around 288.5 eV [145]. Epoxy groups are the main functional groups, while carbonyl groups are the minor functional groups. The ratios of the area of oxygenated peak to sp^2 carbon peak of GQDs-175 °C-1, GQDs-175 °C-2 and GQDs-175 °C-3 are 0.61, 0.47 and 0.33, respectively. All the peaks are centered at 285 nm, which is the result of the charge compensation of the insulator. The area ratio in GQD-175 °C-2 and GQD-175 °C-3 are both higher than GQD-175 °C-1. The tendency is the same with PL intensity, GQD-175 °C-2 is the highest and GQD-175 °C-3 is a little less than GQD-175 °C-2 and GQD-175 °C-1 is the lowest. This result can show the chemical composition change due to the further oxidation by the addition of H_2O_2.

6.2.3 Long-Term Aging Test of GQDs

Figure 6.3a shows the PL spectra of GQD-1, 2 and 3 prepared at 125 °C and 150 °C after 5 days stored at 4 °C. The PL peaks of GQD 125 °C-1 and GQD 150 °C-1 are centered at 504 nm, GQD-150 °C-2, GQD-125 °C-3 and GQD-150 °C-3 are centered at around 490 nm, but GQD-125 °C-2 has a redshift at 518 nm. In Fig. 6.3d and e, after four months, both GQD 125 °C-1 and GQD 150 °C-1 emit yellow-orange color under UV light, while GQD-125 °C-2, GQD-125 °C-3, GQD-150 °C-3 and GQD-150 °C-3 are in blue and cyan color without any changes compared with GQDs just prepared. Moreover, the intensity of GQD-125 °C-3 is the highest, then GQD-125 °C-2 and GQD-150 °C-3 are higher than GQD-150 °C-2, and GQD-125 °C-1 and GQD-150 °C-1 are the lowest. With five days stored at 4 °C, the intensity of GQDs-150 °C does not change much, but a significant increase in intensity is shown for GQDs-125 °C. In Fig. 6.3b the UV–Vis spectra of GQD-125 °C-1, GQD-125 °C-2, GQD-125 °C-3, GQD-150 °C-1, GQD-150 °C-2 and GQD-150 °C-3 after 5 days in the fridge are shown. The typical UV absorption peak of GQD at 230 nm is attributed to the π-π* transition of the aromatic sp^2 domains and the absorption over 300 nm is the n-π* transition [146]. A shoulder around 320 nm observed for as-synthesized GO may be attributed to n–π* transitions of C=O [138]. The UV peak of GQD-125 °C-1 and GQD-150 °C-1 is at around 320 nm, and the peak of GQD-150 °C-2 is at 290 nm, while the peak of GQD-125 °C-2 and GQD-150 °C-3 is at around 270 nm, and GQD-125 °C-3 is at 250 nm. The peak position tendency is same with PLE intensity, which means the higher oxidation level leads to higher PLE intensity. The oxidation level may be the dominant factor to determine the PLE intensity. But the peak position is not the same as this tendency, so other factors should be determined. Figure 6.3c shows the PL spectra of GQDs-175 °C and GQDs-200 °C after one month stored at 4 °C. Among all GQDs, GQD-175 °C-3 has the highest PL intensity. In Fig. 6.3f and g, similar to GQD-125 °C-1 and GQD-150 °C-1, the emission color of GQD-175 °C-1 and GQD-200 °C-1 change to yellow-orange color. In addition, the emission color of GQDs-2 and GQDs-3 are not changing. GQD-175 °C-2, GQD-175 °C-3 and GQD-200 °C-3 emit a similar green–blue color, while GQD-200 °C-2 emits the blue color.

GQDs prepared with adding H_2O_2 show better stability in the aging process. The red-shift for GQD-1 is gradually increasing as time flies. The explanation for red-shifted PL is that the GQDs were strong acid oxidized in the aging process with further cleavage to generate additional oxygenated functional groups, resulting in lower-lying defect states emitting more to the red region in the PL spectrum [94]. However, for GQDs-2, adding H_2O_2 in the strong acidic condition is an exothermic reaction, the hydroxyl radicals ·OH are originated spontaneously [147]. The hydroxyl radicals are very active and further oxidize the remaining sites of GQDs-1. Due to XPS results, GQDs-2 and GQDs-3 generate more epoxy groups and carboxyl groups than GQDs-1. This can be explained by ·OH attacking the remaining carbon bonds by an addition reaction. For epoxy groups, ·OH attacks unsaturated carbon atoms with a positive charge to form carbon-hydroxyl (C–OH) bonds. Then, with concentrated H_2SO_4 as a catalyst, two C–OH have dehydration reaction to form C–O–C groups (C–O–C). For carboxyl groups, ·OH reacts with C-O and transform single bond to double bonds. Compared to GQDs-2 and GQDs-3, with temperature increasing, the degree of oxidation raised progressively and increasing the concentration of GQDs. Therefore, GQDs-3 have higher PL intensity than GQDs-2 but the peak centers hardly ever change.

Moreover, Fig. 6.4 shows post-treatment also influences the PLE intensity of GQD. Figure 6.4a shows the PLE spectra of GQD-125 °C-3 and GQD-150 °C-3 by using 0.02 μm and 0.45 μm filters. For GQD-150 °C-3 using both filters, the PL peaks are

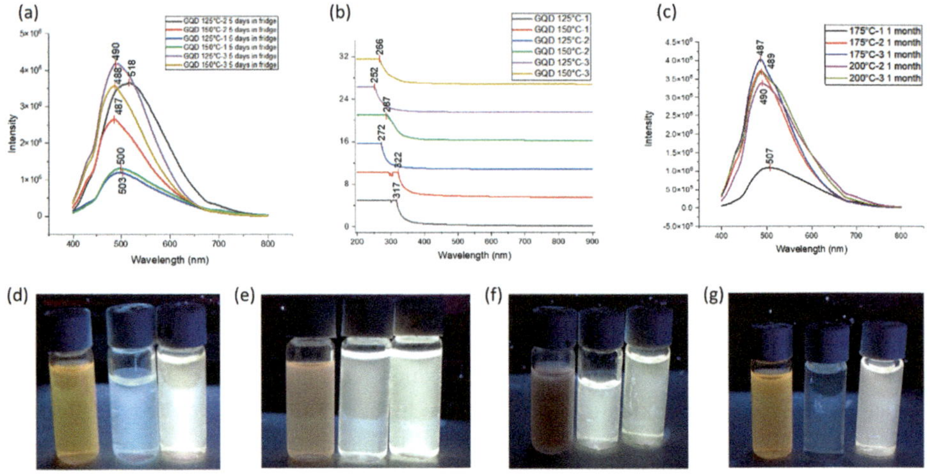

Fig. 6.3 **a** PLE spectra and **b** UV–Vis spectra of GQDs-125 °C and GQDs-150 °C after 5 days at 4 °C, **c** PLE spectra of GQDs-175 °C and 200 °C after one month stored at 4 °C, photographs of GQD stored at 4 °C for four months under UV light, from left to right: **d** GQD-125 °C-1, GQD-125 °C-2 and GQD-125 °C-3, **e** GQD-150 °C-1, GQD-150 °C-2 and GQD-150 °C-3, **f** GQD-175 °C-1, GQD-175 °C-2 and GQD-175 °C-3, **g** GQD-200 °C-1, GQD-200 °C-2 and GQD-200 °C-3

centered at 488 nm, the PLE intensity is almost the same. While for GQD-125 °C-3, both peaks are centered at 490 nm but using a 0.02 μm filter the PL intensity of GQD is higher than that of using a 0.45 μm filter. Therefore, the size of particles between 0.02 and 0.45 μm decreases the PLE intensity. In Fig. 6.4b, PLE of GQD-125 °C-1 and GQD-150 °C-1 with and without centrifuge is shown. In both GQD-125 °C and GQD-150 °C, the PLE intensity of GQD with centrifuge is higher than without centrifuge. The intensity of GQD-125 °C-1 with the centrifuge is almost 2 times of the same sample without a centrifuge. Therefore, post-treatment, including filtration and centrifuge are effectively increases the PLE intensity of GQD. Centrifuge can separate the aggregated GQD in the solution. GQDs can be well-dispersed in the solution for a long time, due to the viscosity in acid solution, the GQDs are hard to be re-aggregation.

6.2.4 Graphitic Structure Analysis

Figure 6.5 displays the Raman spectra of GQD-1, GQD-2 and GQD-3 at 150 °C and 200 °C. In Fig. 6.5a, the Raman spectra of GQD-150 °C-1, GQD-150 °C-2 and GQD-150 °C-3 are presented. The peaks at 1370 cm^{-1} and 1600 cm^{-1} correspond to D-band and G-band, respectively. The D-band represents the defects in the structure, and G-band shows the symmetric vibration of carbon atoms, which is the characteristic peak of graphene-based nanomaterials. The most common G-band is located at 1580 cm^{-1}, so

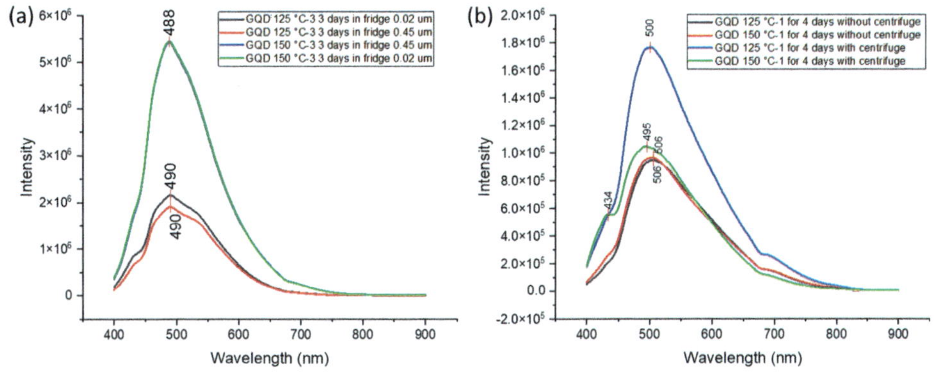

Fig. 6.4 PL emission spectra of GQD with **a** different filter and **b** purification method

the redshift in GQD-150 °C-1 and GQD-150 °C-3 might be attributed to defects and edge structure [148]. The intensity ratio of the D-band and G-band (I_D/I_G) demonstrates the defect density (sp^3/sp^2). The I_D/I_G of GQD-150 °C-1, GQD-150 °C-2 and GQD-150 °C-3 are 1.04, 0.55 and 0.91, respectively. The defects in GQD-150 °C-2 are the lowest, then GQD-150 °C-3 and GQD-150 °C-1. Random distribution of oxygen-containing functional groups on the surface of GQDs interrupt the sp^2 conjugated structure of graphene flakes and form sp^3 clusters, so the value of GQD-150 °C-3 is higher than GQD-150 °C.

The Raman spectra of GQD 200 °C-1, 200 °C-2 and 200 °C-3 are presented in Fig. 6.5b. The peaks at 1370 cm^{-1} and 1600 cm^{-1} show D-band and G-band, respectively, the same peak position as GQD at 150 °C. The intensity ratio of the D-band and G-band (I_D/I_G) demonstrates the defect density (sp^3/sp^2). The I_D/I_G of GQD 200 °C-1, 200 °C-2 and 200 °C-3 are 0.9, 0.45 and 0.72, respectively. The defects in GQD 200 °C-2

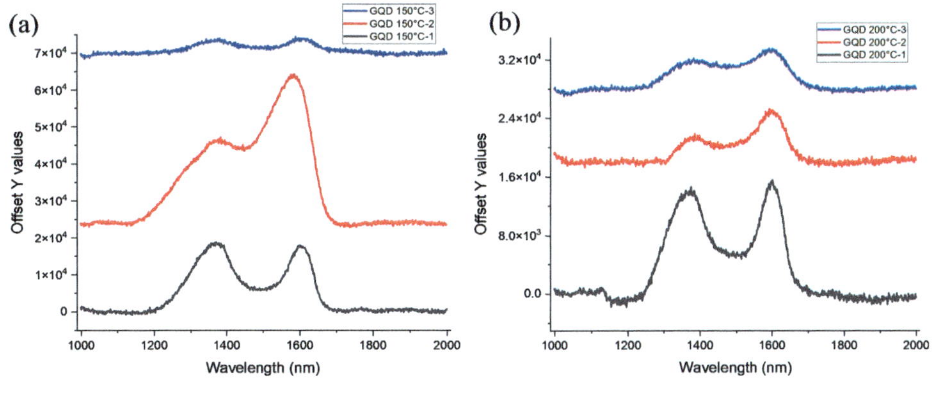

Fig. 6.5 Raman spectra of **a** GQD-150 °C and **b** GQD-200 °C

is the lowest, then GQD 200 °C-3 and GQD 200 °C-1, which shows the same tendency as GQD at 150 °C. The H_2O_2 change both the oxygenated functional groups and the size and number of sp^2 clusters.

6.2.5 Internal Factors Dominating Photoluminescence

The produced GQDs are made of carbon core and surface configurations. The conjugation effect in carbon core, surface state and their synergistic effect should be considered in the PL mechanism. The quantum confinement effect mainly influences the size of GQD. According to Li et al. [149], the PL characteristics of tiny (1.2 nm) CQDs in the UV area (350 nm), medium-sized (1.5–3 nm) CQDs in the visible range (400–700 nm), and big (3.8 nm) CQDs in the near-infrared region (800 nm) changed with particle size. The size of GQDs refers to the area of the sp^2 domain, which is the largest continued conjugated structure rather than the actual size. Due to the quantum confinement effect, for larger-sized GQDs, the bandgap is smaller, so the PLE wavelength increases. In our work, according to PL spectra, the size of GQD is about 1.5 to 3 nm. For GQDs-1, with increasing temperature, the PL wavelength shift from 500 to 505 nm. In Raman spectra, the I_D/I_G of GQD-150 °C-1, GQD-175 °C-1 (Fig. 6.5, I_D/I_G is 0.94) and GQD-200 °C-1 shift from 1.04 to 0.9, also UV–vis absorption wavelength of GQD-125 °C-1 and GQD-150 °C-1 is increased. But in XPS spectra, the ratio of C–O–C to C=C of GQDs-1 at 175 °C is larger than 200 °C. Therefore, with the increasing temperature of strong acid in the cleavage, the size of GQDs is cleaved to smaller but more oxygenated functional groups are added into the structure. After adding H_2O_2, the PL wavelength has a blue shift from 500 to 490 nm except for GQD-125 °C-2 after 5 days of storing in the fridge shift from 500 to 518 nm. Therefore, in high temperatures, H_2O_2 can lead to a blue shift of GQD, while at low temperatures (4 °C for storage), H_2O_2 leads to a redshift. The color range of GQD is from blue to cyan. Compared to GQDs-2 and GQDs-3 at the same temperature, the I_D/I_G of GQDs-2 is always smaller than GQDs-3, and the UV–Vis spectra peak of GQDs-2 and the ratio of functional groups to sp^2 carbon in XPS are always larger than GQDs-3. The oxidation level of GQDs after adding H_2O_2 is higher than acids significantly.

6.3 pH-Controlled Synthesis of GQDs: Tailoring for Device Compatibility

6.3.1 GQD Thin Film Fabrication

For film fabrication, the original solution cannot form film drying in the air due to the nonvolatility of strong acid in GQDs dispersion. The one way to let the film dry was to evaporate the film at 350 °C to let sulfuric acid boil. However, the evaporation leaves white strips on the film and can corrode the electrode materials, such as ITO and Al. Therefore, strong acid should be removed to form the film by neutralization. Sodium hydroxide (NaOH) and Sodium carbonate (Na_2CO_3) can both do the neutralization but are different in the film quality and PL intensity of GQDs.

The first neutralization method was to add NaOH pellets into the GQD dispersion to increase the pH. However, the pellets cannot dissolve in a highly concentrated acid solution and become crystallized. Therefore, NaOH pellets were dissolved in water and neutralized the solution by adding acidic GQD. After the solution was neutralized, it was easy to be crystallized during the film formation. Then, the solution was dialyzed in a dialysis bag (Sigma-Aldrich, MWCO 1 k Da) and put into water. The dialysis time is optimized to be overnight. When it is shortened to 2 h, the film is still crystalized but better than no dialysis. Therefore, longer dialysis can make the film more transparent and have fewer heteroatoms introduction. After dialysis, the concentration of GQDs is less than GQDs in acid. Therefore, evaporation of water at 80 °C increases the concentration of GQDs. Then, the solution can be drop-casting on ITO-coated glass or RCA-cleaned Si substrate and dried in the air. Figure 6.6 shows the optical photos of GQD film neutralized by NaOH with dialysis at 1000 Da for 2 h and overnight. Although with longer dialysis time, salts in the film were easy to be crystalized and form a thick film covering GQDs. Then, the dialysis bag was changed to 2 k Da, so more ions can be exchanged, and the crystallization of the film was lower.

The second method is to add Na_2CO_3 powder to neutralize the acid in the GQD solution. When Na_2CO_3 powder was added to the acid GQD, the solution produced hot gases CO_2. Then, the solution was dialyzed by the dialysis bag at 1 k Da overnight until pH ~ 7 was reached and evaporated to increase the GQD concentration. The film was fabricated by a drop-casting solution on the ITO-coated glass. As in Fig. 6.6c, the GQD film has higher transparency and so thinner thickness than NaOH neutralized film.

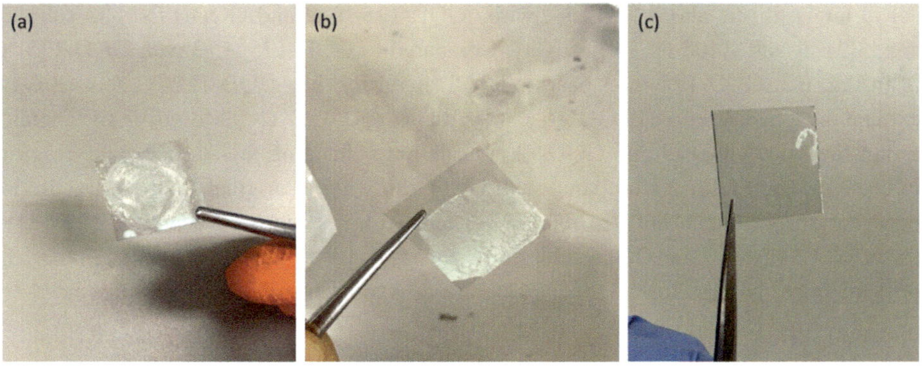

Fig. 6.6 Optical photos of GQD thin-film neutralized by NaOH and **a** dialysis for 2 h, **b** dialysis overnight and **c** GQD film neutralized by Na_2CO_3

6.3.2 Photoluminescence of Neutralized GQD

Figure 6.7 presents the PL spectra of GQD-175 °C-3 and GQD-200 °C-2 solution and films by NaOH or Na_2CO_3 to show the pH-dependent PL properties. In Fig. 6.7a, after neutralization, the PL intensity of the GQD solution is lower, and the PL peak center has a redshift due to GQD concentration and solution pH change. GQDs with carboxyl groups exhibit pH-dependent behavior, which contributes to the protonation/deprotonation of COOH groups in the large difference change in the pH [150]. In the acidic condition, the edge is protonated resulting in broken triplet carbene states, in which luminescence originated from the high concentration of free zigzag sites [87]. In Fig. 6.7b and c, the PL intensity of GQD-175 °C-3 and GQD-200 °C-2 films in a single layer and double layer changes significantly. When depositing GQDs on the ITO-coated glass, the interaction between the GQD solution and the ITO surface made GQD aggregation at the edge. However, when the second layer is deposited, the interaction between the second layer and the top of the first layer is lower, so the GQDs can be evenly distributed. In Fig. 6.7d, the PL intensity of triple-layered GQD film is lower than single-layered due to neutralization by NaOH. With the crystallization of the film, GQDs are distributed uniformly on the substrate. However, with more layers deposited, the film is too thick to block the GQDs absorbing photons. Therefore, the PL intensity of GQDs decrease with the number of layers increased.

6.3.3 Lifetime Analysis

Figure 6.8 compares PL decays of GQD-175 °C-3 film, GQD-175 °C-2 and GQD-175 °C-3 solution. Figure 6.8 (a) shows the decay curve of GQDs, and Fig. 6.8b presents lifetime components extracted from stretched exponential model to do the fitting dialysis. B is the pre-exponential factor, and F is the probability density function (pdf). The lifetime (τ) of GQD film is about 1.242 ns, while GQD-175 °C-2 and GQD-175 °C-3 solutions are 1.391 ns and 1.408 ns, respectively. Compared to GQD-175 °C-2 with GQD-175 °C-3, in Fig. 6.3, GQD-175 °C-3 has higher PL intensity than GQD-175 °C-2, so this can explain the longer lifetime of GQD-175 °C-3 than GQD-175 °C-2. Moreover, the chemical groups between GQDs-3 and GQDs-2 contribute to changing the surface state, resulting in changing the lifetime. Compared with GQD film with its original solution, the lifetime does not change much but distribution for GQD film is more concentrated ($f = 81.48$ for $\tau = 1.242$ ns) than GQD solution ($f = 65.01$ for $\tau = 1.408$ ns). Therefore, the post-treatment process makes the GQDs purified and uniform electronic structure.

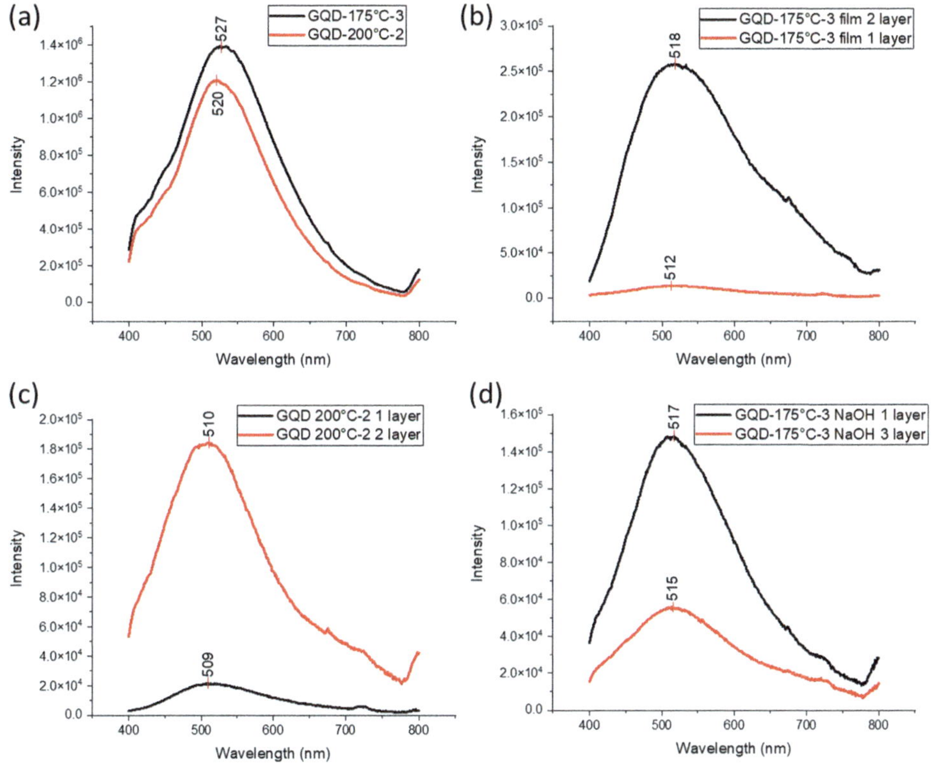

Fig. 6.7 PL spectra of **a** GQD neutralized solution, **b** GQD-175 °C-3, **c** GQD-200 °C-2 film in a single layer and double layers neutralized by Na_2CO_3 and **d** GQD-175 °C-3 film neutralized by NaOH

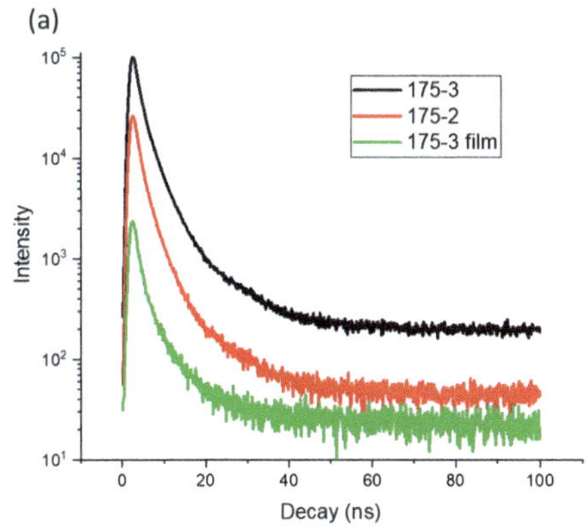

(b)

GQD film					
Peak Num	B	f	τ (ns)	T (ns)	Std τ (ns)
1	0.4586	0.0002	0.1	0.099	0.002
2	16613.61	81.48	1.242	1.164	0.398
3	572.8687	13.81	6.104	6.144	0.824
4	13.1696	4.7098	90.574	98.297	5.947
GQD-175°C-3					
Peak Num	B	f	τ (ns)	T (ns)	Std τ (ns)
1	33.1057	0.0004	0.101	0.099	0.002
2	420000	65.01	1.408	1.384	0.107
3	82027.77	30.14	3.34	3.292	0.178
4	4257.549	4.8504	10.353	10.332	0.281
GQD-175°C-2					
Peak Num	B	f	τ (ns)	T (ns)	Std τ (ns)
1	130000	73.23	1.391	1.337	0.189
2	17579.06	23.07	3.183	3.18	0.249
3	850.512	3.6941	10.534	10.697	0.322

Fig. 6.8 **a** Time-resolved fluorescence decay curves of GQDs and **b** distribution analysis

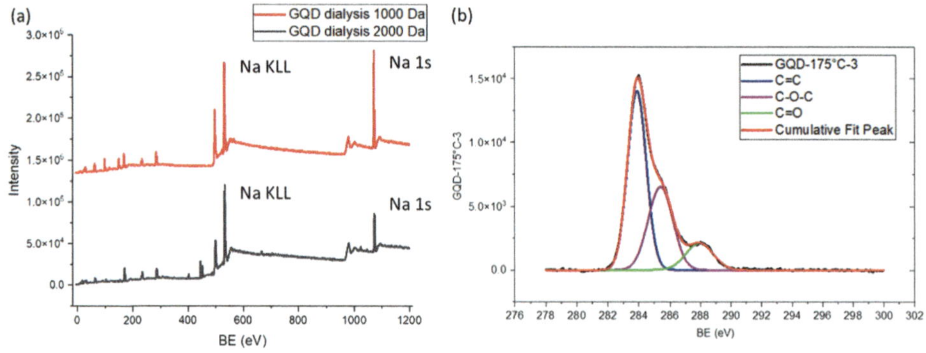

Fig. 6.9 **a** XPS survey spectra of GQD film using dialysis bag with MWCO 1000 Da and 2000 Da and **b** XPS deconvolution of C 1 s of GQD film

6.3.4 Chemical Composition of GQD Film

Figure 6.9 shows the XPS survey spectra and deconvolution of C 1 s of GQD film. Figure 6.9a shows the GQD films neutralized by NaOH and dialyzed by different-sized dialysis bags. The Na 1 s is located at around 1071 eV [151]. The solution inside the dialysis can have balance with water to reduce the concentration of ions. The amount of Na in GQDs dialysis with 2 k Da was lower than 1 k Da. Figure 6.9b shows the deconvolution of C 1 s of GQD-175 °C-3 after the post-treatment process. The chemical composition was characterized as C=C/C–C at a binding energy of 284.8 eV, C–O–C at around 286 eV, and C=O at around 288.5 eV [145]. Compared with Fig. 6.2, the type of functional groups was the same as the original GQD solution but the area percentage of C=C, C–O–C and C=O changed to 0.54, 0.33 and 0.12, respectively. The percentage decrease in C–O–C is due to a ring-opening reaction in alkaline conditions [152]. Therefore, the XPS survey results show the effectiveness of dialysis bags in removing salts and avoiding crystallization of the film. The GQD films have the same oxygen-containing functional groups with modified percentages.

6.4 Summary

In conclusion, the produced GQDs using strong acid and H_2O_2 can have high-intensity multiple color PL and modified pH to fabricate GQD films. The carbon core and the surface state of GQD both influence the PL properties. The effect of H_2O_2 on PL is shown by the change in the intensity and peak wavelength. With different temperatures, the PL wavelength of GQDs can have both blue and redshifts with H_2O_2. The produced GQDs are promising to use in optoelectronic devices.

7 Top-Down Synthesized Graphene Quantum Dots in Photovoltaic Devices

7.1 Concept of the GQDs-Based Photodetectors

QDs are functionalized as absorbers in photodetectors and photovoltaic devices. Absorbing photons with energy larger than the bandgap, the free carriers are generated, and the conductivity is controlled by the absorption, allowing QDs to be applied in light detection. Photodetectors are categorized based upon photocurrent generation mechanism into photoconductive and photovoltaic effects. A photoconductor is to amplify the current under illumination compared with a signal in the dark. In the sensitizing centers, holes (or electrons) are circulated many times through the external circuit before recombination, and electrons (or holes) are trapped in the trap state [153].

The surface engineering of GQDs changed the properties of graphene and carrier transport in photoconductors. With different synthesis methods, the bandgap of graphene is opened, various defect states are introduced into the graphene and changed the electronic [154] and chemical [155] properties. A defect midgap state band (MGB) is introduced into the GQDs structure, also called energy trap states, which was first proposed by Zhang et al. [156]. The origin of MGB can be classified into two parts: off-stoichiometry produced in QDs and oxygen-related functional groups attached to the QDs surface [157]. For colloidal quantum dots (CQDs), the prepared lead chalcogenide PbX (X = S, Se) CQDs is off-stoichiometric with excess Pb located on the polar surface, the charge balance is maintained by undercharging the Pb atoms [157]. Therefore, the trap state emerged. For GQDs, the MGB is caused by surface engineering with oxygenated species [158]. MGB plays an important role in the charge transport before electron-hole recombination. The carriers are trapped during the diffusion, so the carrier mobility is decreased. The traps form a new recombination center for the carriers, so the lifetime of the dominant carriers is decreased [157]. The trap-assisted recombination dominates the current generation, limiting the diffusion length of the charge carriers in the QD films [157]. Figure 7.1 shows

F. Sun, *Synthesis and Device Applications of Graphene Derivatives and Quantum Dots*,
Synthesis Lectures on Green Energy and Technology,
https://doi.org/10.1007/978-3-031-56403-1_7

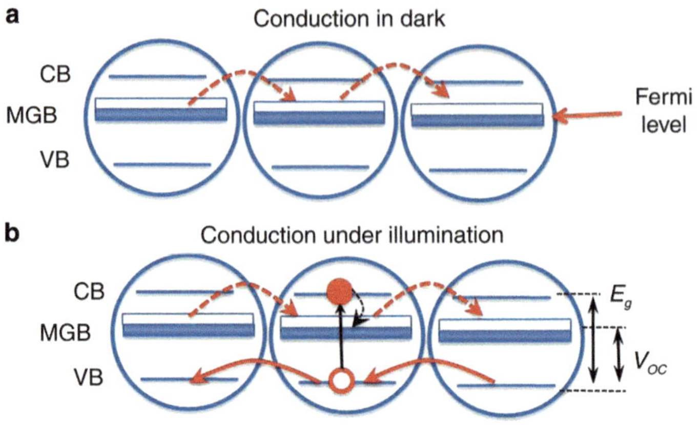

Fig. 7.1 Scheme of conduction mechanism of MGB in QD films in dark and under illumination. Reprinted with permission from [159]. Copyright 2011, Nature

the conduction mechanism of MBG in QD films in dark and under illumination [159]. The MGB in QD films forms a weak conduction band that is full. In the dark, when the negative bias is applied, holes are injected into the MGB, lowering the Fermi level and opening a conducting pathway transport to the next MGB [159]. When the QD films are under illumination, the photogenerated electrons are trapped in MGB from the conduction band (CB) and holes are circulated via the valance band (VB) [159]. MGB acts as recombination centers, trapping both electrons and holes with equal probability and allowing them to be lost by recombination [157].

7.2 Development of Planar GQDs-Based Photodetectors

Figure 7.2 shows the planar device structure, which consists of the top electrode (metal contact), an active layer (GQD), and the bottom electrode (ITO). GQD film is sandwiched between the Al and ITO coated glass. As an active layer in the photoconductive detector, GQD can generate electrons and holes under UV light. When a photon with higher energy than the bandgap energy is absorbed, electron-hole (e-h) pairs are generated, which circulate through an external electric field before recombination [160]. Holes are the charge carriers and electrons trapped in the photoconducstor bulk [161]. ITO is selected to be the electrode due to its high work function [162], electrical conductivity [163], and transparency [162]. The work function of ITO is at around 4.4–4.8 eV, depending on different treatments [162]. The work function of Al is about 4.2 eV [133]. In this structure, holes are transported to ITO and electrons flow towards the metal contact.

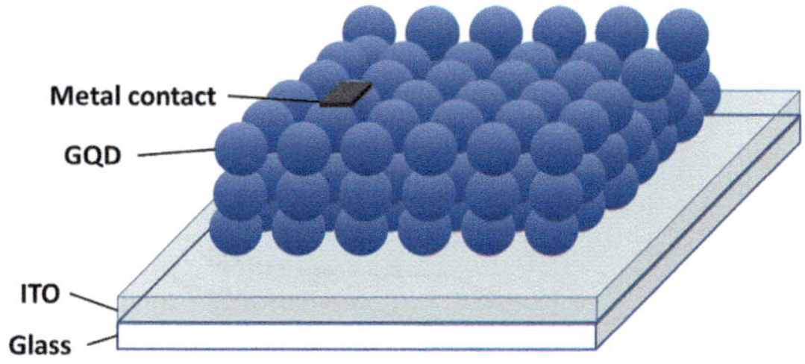

Fig. 7.2 Device scheme of planar GQDs-based photodetectors

7.2.1 Band Structural Analysis for GQD-Based Photodetector

Figure 7.3 shows the cyclic voltammetry (CV) measurement of GQD and band structure of planar GQDs photodetector. In Fig. 7.3 reduction potential (E_{red}) is determined to be −0.6 V, so the lowest unoccupied molecular orbital LUMO) level of GQD is determined as in Eq. 7.1

$$E_{LUMO} = -e(E_{red} + 4.8) = -4.2\,\text{eV} \tag{7.1}$$

The bandgap can be calculated from UV-Vis spectra (Fig. 7.3), which is about 3.1 eV. Therefore, the highest occupied molecular orbital (HOMO) level is about −7.3 eV. Figure 7.3b shows the band structure of the planar GQD photodetector. The GQDs active layer absorbs photons and generates e-h pair, electrons go to Al, which serves as the cathode and holes go to ITO, which serves as the anode. The asymmetric electrode structure can make electrons and holes drift freely to the electrodes. In this work, Ag and Al are used as metal contact, but only Al is shown in the band diagram. The work function of Ag is 4.26 eV, which is similar to Al, so both Ag and Al are acceptable to be the metal contact [103].

7.2.2 Fabrication Process of GQD-Based Photodetectors

The fabrication of the planar device is as follows. The ITO-coated glasses (2.2 × 2.2 cm^2, 5 Ω/sq, Biotain crystal co., limited) were washed with acetone and isopropyl alcohol (IPA) under sonication for 15 min, respectively. Then, the washed glasses were rinsed and dried by a nitrogen gun. A 2.2 × 0.3 cm^2 Kapton tape is attached to the edge of the glass to leave a space to put the probe on the ITO film in the IV measurement. GQD solution was neutralized by Na_2CO_3 and dialyzed by a 1000 Da dialysis bag overnight. 0.5 ml prepared GQD solution was drop-casted on the ITO-coated glass dried in the desiccator overnight. Finally, Al was deposited by e-beam deposition (Nanochrome II, Intlvac Inc., Canada) by using the shadow mask to control the deposition area at around 0.01 cm^2. The final thickness was about 500 nm and the deposition rate is at around 2.5 Å/s.

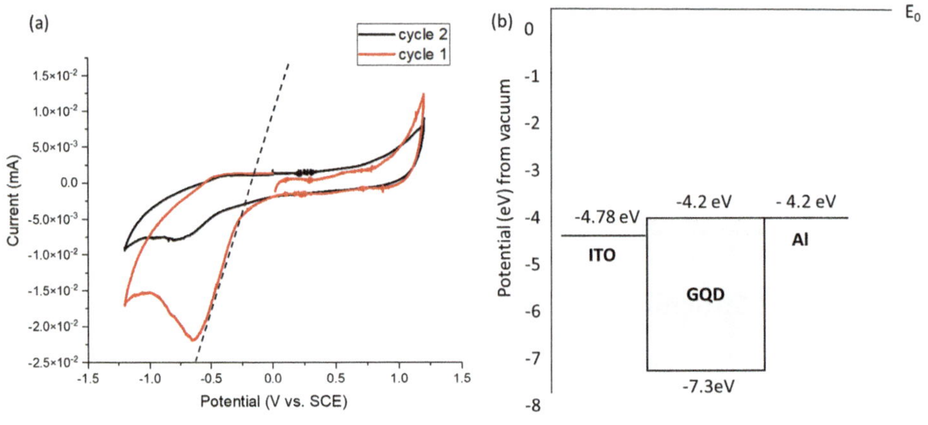

Fig. 7.3 **a** CV measurement of GQD and **b** band diagram of planar GQDs photodetector

7.3 Planar Heterojunction GQD-Based Photodetector

Figure 7.4 shows the device scheme and band diagram of planar heterojunction GQD with MoO_3 as hole transport layer as well as an electron blocking layer. GQD film absorbs photons and generates excitons, electrons move towards the ITO layer and holes travel to the MoO_3 layer then to Al. MoO_3 is utilized as the hole transport layer as well as an electron blocking layer in the device. The HOMO of MoO_3 is at −5.3 eV, which is between the HOMO of GQD of −7.3 eV and the work function of Al of −4.2 eV. The LUMO of MoO_3 is at −2.3 eV, which blocks electrons' transport to the opposite electrode compared to the LUMO of GQD of −4.2 eV.

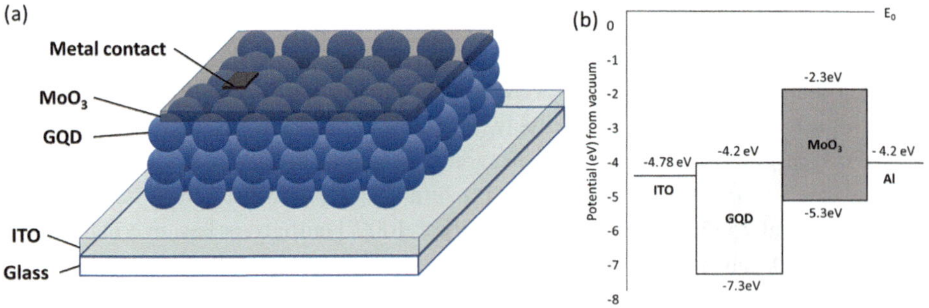

Fig. 7.4 **a** Device scheme and **b** band diagram of planar heterojunction GQD photodetector

7.4 Planar Heterojunction PEG-Passivated GQDs Photodetectors

Figure 7.5 shows the device scheme of planar heterojunction PEG-passivated GQDs with PEDOT: PSS photodetector. In Fig. 7.5a, the device structure is glass/ITO/PEG-passivated GQDs/PEDOT: PSS/Ag. PEG-passivated GQD is functioned as the active layer in the device and forms the heterojunction with PEDOT: PSS. The HOMO of PEDOT: PSS is located at −5.3 eV [164] and its bandgap is about 1.77 eV [165]. PEDOT: PSS shows advantages over MoO_3 for hole extraction. PEDOT: PSS film is fabricated by solution casting method (spin coating) unlike thermal deposition to deposit MoO_3. Moreover, unlike insulating material MoO_3, PEDOT: PSS exhibits high electrical conductivity [166]. Figure 7.5b shows the scheme of formation of GQD with PEDOT: PSS. GQD cross-links with PEDOT through the electrostatic interaction and π-π interaction to form the core-shell nanostructure [164]. The strong interaction between GQD and PEDOT makes the PEDOT chain tightly aligned on the GQD surface via π-π interaction resulting in the improvement of the electrical conductivity of PEDOT: PSS [167]. The formation of hydrophilic groups between GQDs and PSS allows PSS chains to be released and coiled around the edge of GQDs. Due to the heterojunction interface, carrier transport is faster than the double-layered structure and improves the charge separation [167]. Therefore, the planar heterojunction PEG-passivated GQD/PEDOT: PSS can improve the performance of the GQD-based photodetector.

7.4.1 Fabrication of PEG-Passivated GQDs Photodetectors

The planar heterojunction photodetector has the structure of glass/ITO/PEG-GQD/PEDOT: PSS/Ag. For all ITO substrates, a general cleaning procedure was done first. The uncleaned ITO-coated glasses were treated to be more hydrophilic, which involves heating the ITO in a 17:3:1 solution of H_2O:NH_4OH:H_2O_2 for 30 min at 50 °C, rinsed thoroughly with water, and then immersed in acetone for 10 min before being dried with

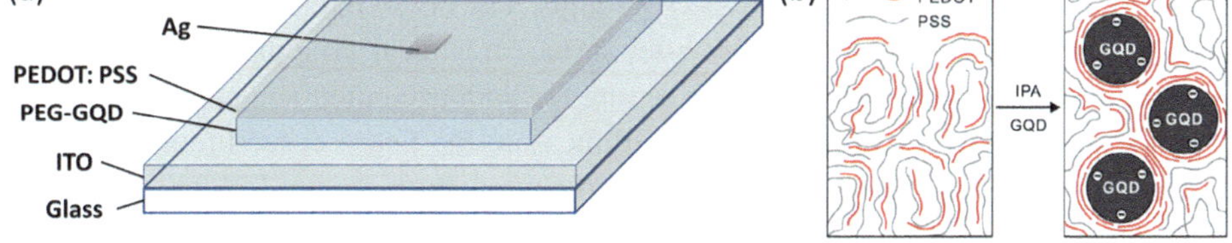

Fig. 7.5 Device scheme of planar heterojunction **a** PEG-passivated GQDs/PEDOT: PSS photodetector and **b** scheme of the formation of the GQD with PEDOT: PSS. Reprinted with permission from [164]. Copyright 2015, ACS Publications

a nitrogen gas stream. Then, PEG-passivated GQD solution was synthesized by mixing prepared GQD and PEG solution (Sigma-Aldrich, 40% (w/w) in H_2O). The ITO-coated glasses (2.2 × 2.2 cm^2, 5 Ω/sq, Biotain crystal co., limited) were applied surface treatment for 30 min. The 5 ml PEG solution was added to 10 ml prepared neutralized GQD dispersed in water at 110 °C by the hotplate for 24 h [168]. The water evaporated completely, so the solvent changed from water to PEG solution. The prepared PEG-GQD solution was spin-coated on ITO-coated glass (500 rpm for 10 s, 3000 rpm for 60 s, and 500 rpm for 15 s). The PEG-GQD film was annealed by a hotplate at 180 °C for 10 min to evaporate the rest solution. Then, PEDOT: PSS solution (Sigma-Aldrich, 1.3 wt% dispersion in H_2O) was spin-coated (500 rpm for 10 s, 3000 rpm for 60 s, and 500 rpm for 15 s) on the PEG-GQD film. The PEDOT: PSS film was annealed by a hotplate at 150 °C for 10 min. Finally, a 30 nm Ag was deposited on the PEDOT: PSS film by thermal evaporation (Nanochrome I, Intlvac Inc., Canada) at the deposition rate of 1–2 Å/s.

7.4.2 Device Characterization Techniques

A scanning electron microscope (SEM) was used to measure the morphology and cross-section of the GQDs-based photodetectors. The GQD samples were prepared by drop-casting on the ITO-coated glass, which was measured by SEM (Zeiss Ultra Plus, Carl Zeiss AG, Germany). The cross-section of the device sample was prepared by cutting the device using the cleavage tool, which was measured by SEM (SU5000, Hitachi, Japan). Both GQD films and the device were sputtered with Au to increase the conductivity before SEM measurement. The PEG-passivated GQD film was fabricated by spin coating on the surface engineered ITO-coated glass, which was measured by SEM (JSM-7200F, JOEL Inc., USA).

Atomic force microscopy (AFM, Dimension Icon AFM, Bruker, USA) was used to investigate the nanoscopic morphology of the surface of the sample. The PEDOT: PSS film was spin-coated on the ITO-coated glass after surface treatment. The mode used in the measurement was auto tapping mode. The dimension of the measured area is 500 × 500 nm^2.

Photoluminescence (PL) spectroscopy (Edinburg Instruments, FLS920, USA) measured the PL emission spectra of PEG-GQD films. The PEG-GQD films were prepared by spin coating on the ITO-coated glass after surface treatment. The excitation wavelength is 360 nm with 8 nm bandwidth in the setting for both excitation and emission.

Time-resolved spectroscopy was used to measure the fluorescence lifetime of GQDs. It includes the fluorescence spectrometer (Edinburgh Instruments, FLS920, USA) with a pulsed LED (Edinburgh Instruments, EPLED-380 USA, 375 nm, 5 M Hz, 943.3 ps pulse width). The fluorescence decay curve was analyzed by the time-correlated single-photon counting (TCSPC) method. The decay data were fitted by the fluorescence analysis software (FAST Version 3.0., Edinburgh Instruments Ltd., UK). PEG-GQD films were prepared by spin-coating the solution on the ITO-coated glass.

UV-Vis spectroscopy (Shimadzu, UV-2501PC, Japanese) investigated the absorption of the sample combining with PL to determine the size. The optical transmittance of the films was measured in the wavelength range of 200–900 nm. The PEG-GQD films were prepared by spin coating on the ITO-coated glass.

Electrometer/high resistance meter (Keysight, B2985A, USA) connected to a probe station was used to measure the current-voltage (IV) curve of the photodetectors in the dark and under illumination by the Xenon Research Arc Lamp Sources (Oriel® Instruments—Newport, Newport 67,005, USA). The power of the lamp is ranged from 50 to 500 W with lens transmittance ranging from 200 to 2500 nm. The power was set to 80 W in the measurement. One 20 μm probe contacts the ITO film and the other touches the metal contact on the top of the device.

The I-V behavior of the device is also measured by a semiconductor Parameter Analyzer (Agilent 4155C, Agilent Technologies, Canada) connected to a 2-probe station inside a dark, sealed chamber. A 4-W UV lamp (Cole-Parmer Handheld Ultraviolet Lamps, Antylia Scientific, USA) with dual 365 nm wavelength tubes was mounted over the device during the measurement. A power supply (Agilent E3620 A, Agilent Technologies, Canada) was used to apply the bias under the test.

Cyclic voltammetry (C-V) (CHI600E electrochemical analyzer) investigated the LUMO (lowest unoccupied molecular orbital) and HOMO (highest occupied molecular orbital) of the sample. The setup of CV consists of Ag/AgCl reference electrodes, two Pt electrodes as working and counter electrodes, respectively, and 0.1 M tetrabutylammonium hexafluorophosphate solution in acetonitrile. In the measurement, the scan rate of 100 mV/s. The reference is Ferrocene, which has a HOMO level of −4.8 eV.

7.5 Results and Discussion of GQDs-Based Photodetectors

7.5.1 Optical Properties of PEG-Passivated GQDs Film

Figure 7.6 shows the UV-Vis absorption and PL emission spectrum of PEG-GQD/PEDOT: PSS film and the scheme of the PL mechanism of GQD. In Fig. 7.6a, the bandgap of PEG-GQD can be calculated from the absorption. By calculating the intercept of the extrapolation of the absorption edge to the x-axis, the absorption starts at around 300 nm [103]. As a result, the optical density is, as an approximate estimate, the bandgap is about 4.1 eV. Figure 7.6b, c show the PL emission spectrum agrees well with the theoretical simulation of the PL mechanism of GQD. The PL emission shows the spin-coated film with high intensity at around 2E+5. The oxygenated functional groups appear to have two functions: one is self-orbitals of functional groups (n, π, σ), which can contribute emissive components with fixed positions on the PL spectrum via electronic transitions

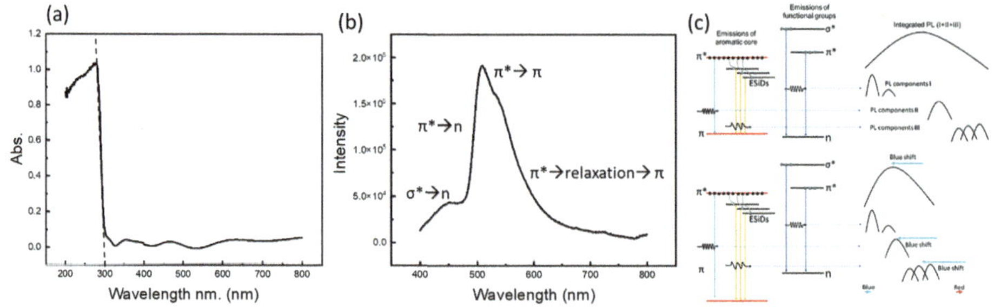

Fig. 7.6 **a** UV-Vis spectrum of PEG-GQD/PEDOT: PSS film, **b** PL emission spectra of PEG-GQD/PEDOT: PSS film, and **c** scheme of PL mechanism of GQD aromatic core and oxygenated function groups emission. Reprinted with permission from [169]. Copyright 2015, Royal Society of Chemistry

from anti-bonding states (π^*, σ^*) to non-bonding (n) states; the other is to cause structural deformation to GQDs [169]. The aromatic core contributes to the PL through $\pi^* \rightarrow$ ESiD $\rightarrow \pi$ and $\pi^* \rightarrow \pi$, and the intensity of PL is depend on the size due to quantum confinement, smaller size, larger bandgap, and higher emission intensity [169]. Functional groups contribute to the PL spectrum through $\pi^* \rightarrow$ n and $\sigma^* \rightarrow$ n [169].

7.5.2 Photodetector Devices Morphology

Figure 7.7 shows the cross-sectional SEM images of GQD film and heterojunction photodetector, SEM image of PEG-passivated GQD film, and AFM topographic image of PEDOT: PSS film. Figure 7.7a shows the cross-section of the drop cast GQD film. The GQD film forms the layered structure during the drying process because the desiccant's drying and solvent evaporation give different pressure on the films. The drop-casting technique produces a thick film with a thickness of around 1–2 μm. This layered structure provides large-scale surface contact. Figure 7.7b shows the thickness of each layer of the photodetector. The spin-coated PEDOT: PSS is about 600 nm and the drop-casted GQD film is about 2.38 μm. The highly stacking structure of PEDOT: PSS exhibits a good film-forming capability. Figure 7.7c presents the morphology of the PEG-passivated GQD film. Compared to GQD film, PEG-GQD film inhibits agglomeration due to the long tail forming the spatial resistance to water molecules resulting in a strong repulsion [170]. The functionalized GQD reduces the π-π stacking in the bare GQD due to the longer chain attached to the surface of the GQD increasing the distance between the two-layer [171]. Surface functionalization increases the solubility, dispersion, and decreases defects in luminescent materials, all of which help to improve GQD stability and fluorescence efficiency [170]. Figure 7.7d shows the topography of the PEDOT: PSS surface. The spin-coated PEDOT: PSS film has good uniformity. The bright and dark sides are the

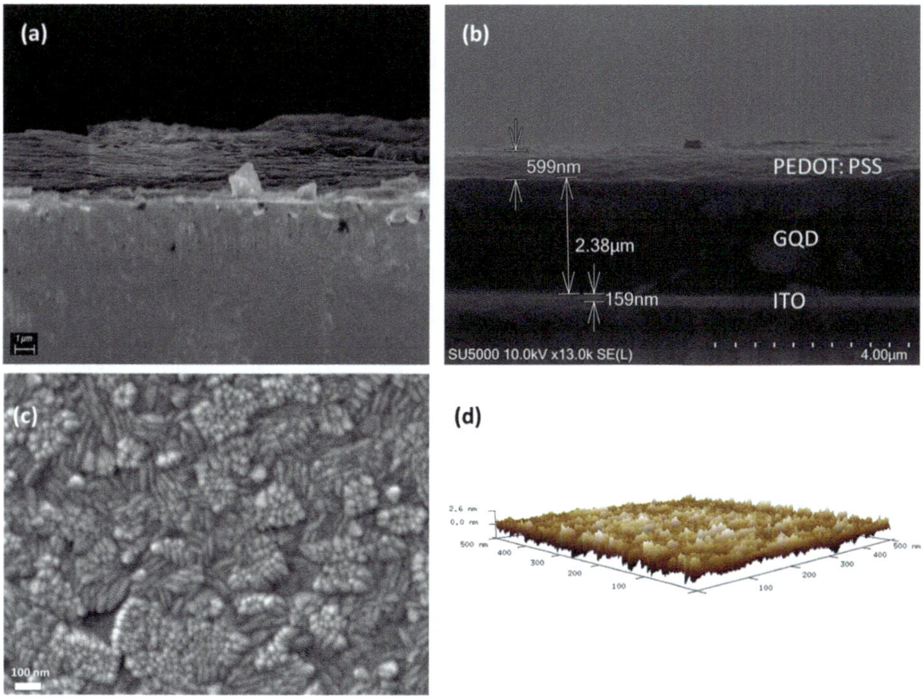

Fig. 7.7 Cross-sectional SEM image of **a** GQD film and **b** GQDs-based heterojunction photodetector device, **c** SEM image of PEG-passivated GQD film, and **d** AFM topography of the surface of PEDOT: PSS film

PEDOT domain and PSS domain which is relatively homogeneous and interconnected so the electrical conductivity is high [164].

7.5.3 Carrier Transport Mechanism

Figure 7.8 shows the PL decay curve of GQD and PEG-GQD solution. The lifetime of PEG-GQD-175°C-3, PEG-GQD-125°C-3, and GQD-175°C-3 is 1.545, 1.734, and 1.242 ns. After being treated with PEG, the lifetime of GQD increased significantly (around 20%). This might be because PEG modifies the conjugated aromatic structure of GQDs, generating an electron delocalization field by combining the PEG ligand's LUMO energy levels with the conductive band-edge energy of GQDs and suppling the excited electrons of PEG-GQDs with a more stable condition, increasing the fluorescence lifetime of PEG-GQDs [170]. Functional groups or disordered structures cause structural deformation, resulting in irregular hybridized π orbitals with mid-gap states (ESiDs) in between the limited π-π^* gap of the aromatic core of dots. These orbitals may act as

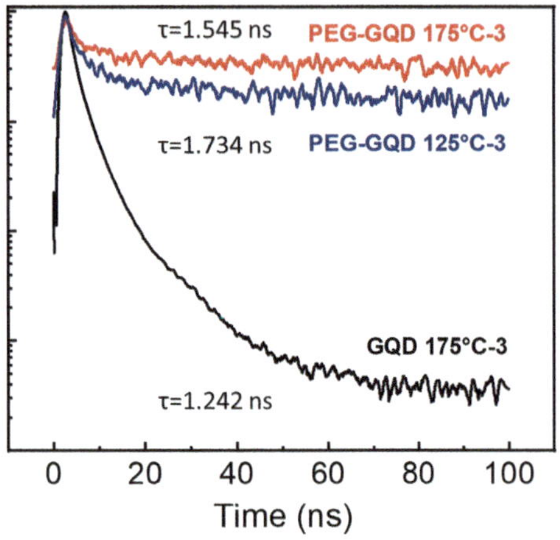

Fig. 7.8 PL decay of GQD and PEG-GQD solution

intermediary relaxation orbitals for electrons transitioning from antibonding states (π^*) to bonding states (π) in the aromatic core of GQDs, contributing to the red side of the broadened emission with a longer PL lifetime [169]. The longer lifetime gives the carrier more opportunity to transport to the electrodes before recombination.

MGB is important in charge transport, especially when the diffusion of excitons or minority carriers dominates the transport. Due to their function as recombination centers, traps reduce both effective carrier mobility and excited-state lifetime. The trap-assisted recombination dominates the diode current, which is the average distance among deep recombination centers that restricts the diffusion length of excited photocarriers in today's CQD films [157]. With functionalization, the lifetime of GQD is prolonged, resulting in mitigating the trap sites and improving the QD film adhesion. Reducing MGB further enhances diffusion length, minimizes recombination and trapping, and improves the performance of the device. The interface contact resistance is decreased to improve the realization of charge separation.

7.5.4 Current-Voltage (I-V) Response to UV Exposure

Figure 7.9 analyzes the I-V behavior of the devices under dark and UV conditions. Dark current was measured under the sealed black chamber by turning off the UV lamp mounted over the device. Then, the UV lamp was switched to exposure UV light on the device for 5 min until the current stopped increasing. When applied to -1 eV, the dark and UV current is 2.5E$-$5 A and 1.1E$-$4 A, respectively. When applied to 1 eV, the dark and UV current is 2.2E$-$5 A and 8.4E$-$5 A, respectively. The photo-to-dark current (ON/OFF) rate is about 400%. Due to UV light being the only illumination, the significant increase of the current comes from UV exposure. Due to MGB, the presence of trap states and sensitization centers in this planar structural device extends the lifetime of electrons

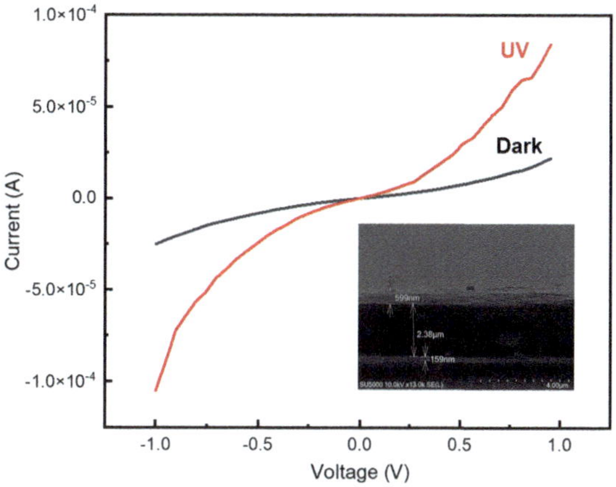

Fig. 7.9 I-V behavior of heterojunction GQD/PEDOT: PSS photodetector

while allowing the holes to be circulated several times in the external circuit, resulting in a significant photoconductive gain [172]. UV exposure may also improve the PL of GQD by converting −OH groups to −COOH groups, which enhance the surface state and change the lifetime [173]. Under UV exposure longer, the shorter lifetime belongs to the non-radiative process [93].

Figure 7.10 shows the dark current, after UV exposure and dark current again (turn off the UV lamp). The second dark current confirmed the current change came from the UV exposure. The device structure is glass/ITO/GQD/MoO_3/Al. The neutralized GQD is dispersed in ethanol instead of water to improve the adhesion between the GQDs and ITO-coated glass. Moreover, due to the smaller contact angle of ethanol than water on the ITO-coated glass, the thickness of drop-casted GQD film is thinner than water-based [174]. The single-layered GQD was fabricated by drop-casting GQD solution on the ITO-coated glass and dried in the desiccator overnight. The second layer was dropped on the first layer and dried overnight. Then, the double-layered GQD was annealed by the hotplate at 150 °C for 10 min. Without annealing for the first layer, the two layers can be mixed with less contact resistance. In Fig. 7.10a, when applied −2 eV, the first dark current was at 8.9E−8 A, after UV exposure, the current increased to 1.8E−7 A and the second dark current went back to 6E−8 A. In Fig. 7.10b, the dark current was at 9.5E−6, then, under UV exposure the current increased to 6.7E−5 A, and then turn off the UV lamp, the dark current went back to 1.3E−5 A. The dark current increased due to higher free carrier density, which comes from the higher concentration of the GQD layer. The ON/OFF rate also increased from 200 to 500% after adding the new GQD layer.

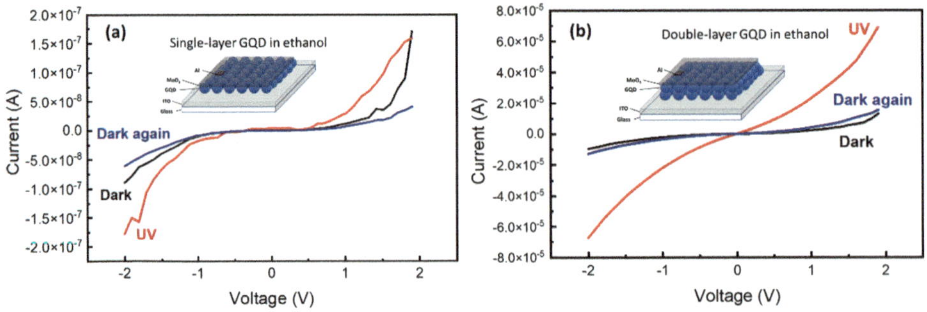

Fig. 7.10 I-V behavior of heterojunction GQD/MoO_3 photodetector with **a** single-layer GQD film and **b** double-layer GQD film

7.5.5 Current-Voltage (I-V) Response to Visible Illumination

Figure 7.11 analyzes the I-V behavior of the devices under dark and illumination conditions. The illumination source is the 80-W Xenon lamp with a 200–2500 nm wavelength. The device structure is glass/ITO/GQD/PEDOT: PSS/Ag. The GQD film was drop cast on the ITO-coated glass and dried in the desiccator overnight. Then, PEDOT: PSS film was spin-coated on the GQD film and annealed using a hotplate at 150 °C for 10 min. In Fig. 7.11a, the current is measured in the loop to confirm the current change came from the illumination instead of current fluctuation. The device shows a diode-type I-V curve. After the illumination, the current increased from 1E−5 A to 2E−5 A. Compared to MoO_3, the spin-coated PEDOT: PSS film is much thicker but exhibits better conductivity resulting in a higher dark and illuminated current and lower ON/OFF rate at 200%. In Fig. 7.11b, the PEG-GQD film was spin-coated on the ITO-coated glass and annealed at 200 °C for 10 min to evaporate the rest solvent. Then, PEDOT: PSS was spin-coated on the PEG-GQD film and annealed at 150 °C for 10 min. The high-power illumination source may provide the annealing effect on PEG-GQDs to change the ligand structure and interdot spacing [175].

7.6 Summary

A planar structured GQD-based heterojunction photodetector was fabricated by using GQDs as the active layer and MoO_3 or PEDOT: PSS as the hole transport layer. The I-V behavior was studied from the mechanism of carrier transport and materials properties. The mechanism of the GQD-based photodetector is explained by MGB in the band structure. With the functionalization of GQD, the optical and electronic properties changed. In the photodetector, the hole transport layer using inorganic or organic material shows the

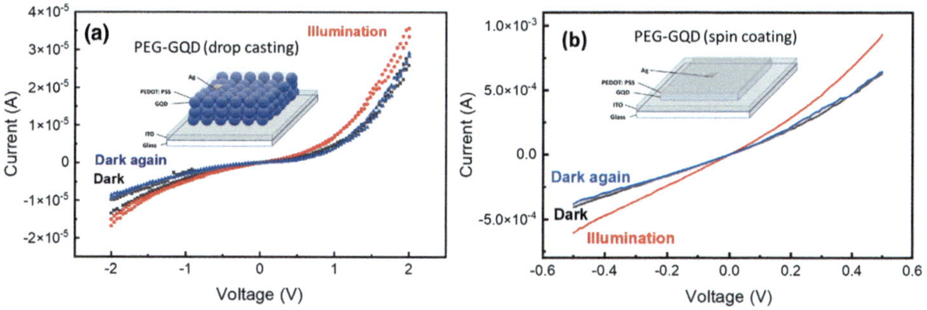

Fig. 7.11 I-V behavior of heterojunction **a** GQD and **b** PEG-passivated GQD photodetector

difference in the dark and illuminated current. This chapter presents the process developed and the results of the heterojunction photodetector.

8 Conclusions and Outlook

8.1 Conclusions

In this work, alternative graphene-based materials synthesis, and the development of fabrication of their applications in electronic devices were presented. First, the novel, high-throughput chemical synthesis approaches for graphene-based products were developed. Then, the materials were optimized to be applied in photovoltaic and photodetector devices. Graphene -based materials were synthesized environmentally friendly, and exhibiting high performance in electrical properties, optical properties, and chemical stability. RGO thin film had high transparency which can be comparable with ITO. Work function of RGO can be tuned in a large range from 3.6 to 5.9 eV by controlling the oxidation degree. GQD was synthesized by improved top-down method with high PL intensity and broad emission range. The prepared materials were modified to increase the compatibility between interfaces and applied in the optoelectronic devices.

The precursor GO was synthesized by the modified Hummer's method with no emission of toxic gas, such as NO_x due to elimination of $NaNO_3$ in the synthesis. The production yield reached 50% as reported in Chapter 3. The produced GO has oxygenated functional groups attached to the surface and edge, leading to high hydrophilicity properties. During the oxidation, the size of graphite was expanded by oxygenated functional groups attached to the planar surface. Moreover, the multi-layered thin film was fabricated by spin coating with high uniformity. Larger size and uniform film paved the way for GO to be reduced to RGO in the next step.

RGO was produced from the prepared GO by a mild and novel two-step method using L-AA and HI as the reductants to improve the electrical conductivity and tune the work function. The produced RGO thin film with a thickness of 20 nm exhibited high electrical conductivity in the range of 360–480 S/m, and high transparency at 90% or higher in the range of 450–800 nm. The uniform RGO films were fabricated from GO thin

F. Sun, *Synthesis and Device Applications of Graphene Derivatives and Quantum Dots*,
Synthesis Lectures on Green Energy and Technology,
https://doi.org/10.1007/978-3-031-56403-1_8

films by decreasing the defects to restore the π-π conjugation structure and eliminate the oxygen-containing functional groups attached. The structure change not only influenced the electrical conductivity but also adjusted the work function. From GO to RGO, the GO was first reduced by L-AA to form the intermediate RGO, then the final RGO was continuously produced by using HI as the reductant. With the different levels of reduction degree, the work function of GO to RGO was in the range of 3.6–5.9 eV. Therefore, the reduction process can effectively control the electronic properties of RGO.

The prepared RGO was applied to MOSCAP devices as the top gate metal with the high performance of C-V measurement. A simple MOSCAP device with RGO as the gate metal was fabricated with the optimized process to deposit top contact (RGO), an oxide layer (SiO_2) and back contact (Al). The C-V curve of the MOSCAP device showed accumulation, depletion, and inversion regions. To enhance the electrical conductivity and decrease the work function difference between the metal and oxide layer, the produced RGO was produced by thermal annealing reduction. With the high C/O, RGO showed high electrical conductivity at 5000 S/m and effective work function at 4.7 eV. The findings of this research are not confined to MOS technology; they may be used in other technologies such as OLEDs and OPVs in which graphene is offered as a conducting electrode.

The GQD was produced from graphite as the starting material by using the modified top-down method. The graphite was oxidatively first cut with strong acid and continuously cut by adding H_2O_2. The enhancement after the second and third step cutting, the PL intensity increased with a shift in the PL peaks to emit multiple color. PL spectra of GQD were influenced by the carbon core and surface states. H_2O_2 produced free radicals to improve the oxidation level and cut more sites in the graphite sheets to modify the core size and surface functional groups. However, the solution with strong acidity was hindered by the fabrication of GQD thin film. The modified pH process by neutralization of the solution was applied to the GQD solution. The GQD solvent was changed to water and can be drop-casting to form the films with high-intensity PL spectra. The present research provides the enhancement in the PL intensity of GQDs in the wide range of PL peak wavelength and facilities the produced GQD for optoelectronic devices.

The produced GQD was applied as the active layer in the GQD-based heterojunction photodetector device with a hole transport layer. The hole transport layer compared MoO_3 and PEDOT: PSS to optimize the work function difference and enhance the photocurrent. The mechanism of the GQD-based photodetector is based on MGB theory to explain the carrier transport and recombination rate. To further enhance the photocurrent generation, GQD was functionalized by PEG to change the length of the ligand and the viscosity of the solution. GQD film was fabricated by spin coating to be controlled at about 100 nm. Therefore, the ON/OFF rate was further improved to make the high-performance photodetector devices.

8.2 Outlook and Future Work

This section contains suggestions for expanding on the work described in this book. Material quality, film quality, process adaptability, and device performance will all be enhanced due to these investigations.

Compared with other graphene synthesis methods, such as CVD, epitaxial growth on SiC, liquid-phase exfoliation, oxidation–reduction method can produce higher yield, lower cost GO but with higher defects density, lower purity, thickness uniformity and smaller lateral size [176]. To further the industrialization of graphene, GO synthesis should improve the tight control over the deposition of a known number of graphene layers with a defined quantity of oxygen concentration.

The large area of GQD interconnecting with solids gives opportunities and challenges to the applications. With the wide range of materials attached to the surface, surface states as the dominant role can be required advances in synthesis method, experimental procedure, and simulation of the control of the ambient environment to be applied in the different devices [153]. The interface between QD and other materials should be matched, introducing contact resistance conducts the performance of the devices. Therefore, surface engineering on the QD layer and other components plays an important role to optimize the device performance [153].

Bibliography

1. Gao XJ, Elowitz MB (2016) Twenty-five years of low-cost solar cells. Nature 538:463–464
2. DeWolf S, Descoeudres A, Holman ZC, Ballif C (2012) High-efficiency silicon heterojunction solar cells: a review. Green 2:7–24. https://doi.org/10.1515/green-2011-0018
3. Lee TD, Ebong AU (2017) A review of thin film solar cell technologies and challenges. Renew Sustain Energy Rev 70:1286–1297. https://doi.org/10.1016/j.rser.2016.12.028
4. Morales-Acevedo A (2006) Thin film CdS/CdTe solar cells: research perspectives. Sol Energy 80:675–681. https://doi.org/10.1016/j.solener.2005.10.008
5. NREL (2019) Best research-cell efficiency Chart.pdf. Nrel 1
6. Ren W, Cheng HM (2014) The global growth of graphene. Nat Nanotechnol 9:726–730. https://doi.org/10.1038/nnano.2014.229
7. Zhong Y, Zhen Z, Zhu H (2017) Graphene: fundamental research and potential applications. FlatChem 4:20–32. https://doi.org/10.1016/j.flatc.2017.06.008
8. Ohta T, Bostwick A, Seyller T, Horn K, Rotenberg E (2006) Controlling the electronic structure of bilayer graphene. Science (1979) 313:951–954. https://doi.org/10.1126/science.1130681
9. Chen H, Müller MB, Gilmore KJ, Wallace GG, Li D (2008) Mechanically strong, electrically conductive, and biocompatible graphene paper. Adv Mater 20:3557–3561. https://doi.org/10.1002/adma.200800757
10. Bolotin KI, Sikes KJ, Jiang Z, Klima M, Fudenberg G, Hone J, Kim P, Stormer HL (2008) Ultrahigh electron mobility in suspended graphene. Solid State Commun 146:351–355. https://doi.org/10.1016/j.ssc.2008.02.024
11. Balandin AA, Ghosh S, Bao W, Calizo I, Teweldebrhan D, Miao F, Lau CN (2008) Superior thermal conductivity of single-layer graphene. Nano Lett 8:902–907. https://doi.org/10.1021/nl0731872
12. Nair RR, Blake P, Grigorenko AN, Novoselov KS, Booth TJ, Stauber T, Peres NMR, Geim AK (2008) Fine structure constant defines visual transparency of graphene. Science (1979) 320:1308. https://doi.org/10.1126/science.1156965
13. Wang W, Shen C, Li S, Min J, Yi C (2014) Mechanical properties of single layer graphene nanoribbons through bending experimental simulations. AIP Adv 4. https://doi.org/10.1063/1.4868625

F. Sun, *Synthesis and Device Applications of Graphene Derivatives and Quantum Dots*, Synthesis Lectures on Green Energy and Technology,
https://doi.org/10.1007/978-3-031-56403-1

14. Novoselov KS, Geim AK, Morozov SV, Jiang D, Zhang Y, Dubonos SV, Grigorieva IV, Firsov AA (2004) Electric field effect in atomically thin carbon films. 306:666–669
15. Berger C, Wu X, Brown N, Naud C, Li X, Song Z, Mayou D, Li T, Hass J, Marchenkov A, Conrad EH, First PN, De Heer WA (2006) Electronic confinement and coherence in patterned epitaxial graphene. Science (1979) 312:1191–1196
16. Li X, Cai W, An J, Kim S, Nah J, Yang D, Piner R, Velamakanni A, Jung I, Tutuc E, Banerjee SK, Colombo L, Ruoff RS (2009) Large-area synthesis of high-quality and uniform graphene films on copper foils. Science (1979) 324:1312–1314. https://doi.org/10.1126/science.1171245
17. Dikin DA, Stankovich S, Zimney EJ, Piner RD, Dommett GHB, Evmenenko G, Nguyen ST, Ruoff RS (2007) Preparation and characterization of graphene oxide paper. Nature 448:457–460. https://doi.org/10.1038/nature06016
18. Sun Z, Yan Z, Yao J, Beitler E, Zhu Y, Tour JM (2010) Growth of graphene from solid carbon sources. Nature 468:549–552. https://doi.org/10.1038/nature09579
19. Chen D, Feng H, Li J (2012) Graphene oxide: preparation, functionalization, and electrochemical applications. Chem Rev 112:6027–6053. https://doi.org/10.1021/cr300115g
20. Lerf A, He H, Forster M, Klinowski J (1998) Structure of graphite oxide revisited. J Phys Chem B 102:4477–4482. https://doi.org/10.1021/jp9731821
21. Dreyer DR, Todd AD, Bielawski CW (2014) Harnessing the chemistry of graphene oxide. Chem Soc Rev 43:5288–5301. https://doi.org/10.1039/C4CS00060A
22. Mativetsky JM, Treossi E, Orgiu E, Melucci M, Veronese GP, Samorì P, Palermo V (2010) Local current mapping and patterning of reduced graphene oxide. J Am Chem Soc 132:14130–14136. https://doi.org/10.1021/ja104567f
23. Gómez-Navarro C, Weitz RT, Bittner AM, Scolari M, Mews A, Burghard M, Kern K (2007) Electronic transport properties of individual chemically reduced graphene oxide sheets. Nano Lett 7:3499–3503. https://doi.org/10.1021/nl072090c
24. Erickson K, Erni R, Lee Z, Alem N, Gannett W, Zettl A (2010) Determination of the local chemical structure of graphene oxide and reduced graphene oxide. Adv Mater 22:4467–4472. https://doi.org/10.1002/adma.201000732
25. Kudin KN, Ozbas B, Schniepp HC, Prud'homme RK, Aksay IA, Car R (2008) Raman spectra of graphite oxide and functionalized graphene sheets. Nano Lett 8:36–41. https://doi.org/10.1021/nl071822y
26. Al-Gaashani R, Najjar A, Zakaria Y, Mansour S, Atieh MA (2019) XPS and structural studies of high quality graphene oxide and reduced graphene oxide prepared by different chemical oxidation methods. Ceram Int 45:14439–14448. https://doi.org/10.1016/j.ceramint.2019.04.165
27. Guerrero-Contreras J, Caballero-Briones F (2015) Graphene oxide powders with different oxidation degree, prepared by synthesis variations of the Hummers method. Mater Chem Phys 153:209–220. https://doi.org/10.1016/j.matchemphys.2015.01.005
28. Becerril HA, Mao J, Liu Z, Stoltenberg RM, Bao Z, Chen Y (2008) Evaluation of solution-processed reduced graphene oxide films as transparent conductors. ACS Nano 2:463–470. https://doi.org/10.1021/nn700375n
29. Loh KP, Bao Q, Eda G, Chhowalla M (2010) Graphene oxide as a chemically tunable platform for optical applications. Nat Chem 2:1015–1024. https://doi.org/10.1038/nchem.907
30. Wang M, Niu Y, Zhou J, Wen H, Zhang Z, Luo D, Gao D, Yang J, Liang D, Li Y (2016) The dispersion and aggregation of graphene oxide in aqueous media. Nanoscale 8:14587–14592. https://doi.org/10.1039/c6nr03503e
31. May R, May R (1859) On the atomic weight of graphite. JBy

32. Salze D, Doppel D, Reaction D, Staudenmaier L, Substanz D, Zeit I (1898) Verfahren zur Darstellung der Graphitsaure. Berichte der Dtsch Chem Gesellschaft
33. Hummers WS, Offeman RE (1958) Preparation of graphitic oxide. J Am Chem Soc 80:1339. https://doi.org/10.1021/ja01539a017
34. Kovtyukhova NI, Ollivier PJ, Martin BR, Mallouk TE, Buzaneva EV, Gorchinskiy AD (1999) Layer-by-layer assembly of ultrathin composite films from micron-sized graphite oxide sheets and polycations. Chem Mater 11:771–778. https://doi.org/10.1021/cm981085u
35. Marcano DC, Kosynkin DV, Berlin JM, Sinitskii A, Sun Z, Slesarev A, Alemany LB, Lu W, Tour JM (2010) Improved synthesis of graphene oxide. ACS Nano 4:4806–4814. https://doi.org/10.1021/nn1006368
36. Chen J, Yao B, Li C, Shi G (2013) An improved Hummers method for eco-friendly synthesis of graphene oxide. Carbon N Y 64:225–229. https://doi.org/10.1016/j.carbon.2013.07.055
37. Peng L, Xu Z, Liu Z, Wei Y, Sun H, Li Z, Zhao X, Gao C (2015) An iron-based green approach to 1-h production of single-layer graphene oxide. Nat Commun 6:1–9. https://doi.org/10.1038/ncomms6716
38. Yu C, Wang CF, Chen S (2016) Facile access to graphene oxide from ferro-induced oxidation. Sci Rep 6:1–7. https://doi.org/10.1038/srep17071
39. Dimiev AM, Tour JM, Science M, Science C, Science N, Street M, States U, Materials AZE, Avenue M, States U (2014) Mechanism of graphene oxide formation. ACS Nano 3060–3068. https://doi.org/10.1021/nn500606a
40. Eigler S, Dotzer C, Hof F, Bauer W, Hirsch A (2013) Sulfur species in graphene oxide. Chem Eur J 19:9490–9496. https://doi.org/10.1002/chem.201300387
41. Chen J, Zhang Y, Zhang M, Yao B, Li Y, Huang L, Li C, Shi G (2016) Water-enhanced oxidation of graphite to graphene oxide with controlled species of oxygenated groups. Chem Sci 7:1874–1881. https://doi.org/10.1039/c5sc03828f
42. Pendolino F, Armata N, Masullo T, Cuttitta A (2015) Temperature influence on the synthesis of pristine graphene oxide and graphite oxide. Mater Chem Phys 164:71–77. https://doi.org/10.1016/j.matchemphys.2015.08.024
43. Sun L, Fugetsu B (2013) Mass production of graphene oxide from expanded graphite. Mater Lett 109:207–210. https://doi.org/10.1016/j.matlet.2013.07.072
44. Pei S, Cheng H-M (2012) The reduction of graphene oxide. Carbon N Y 50:3210–3228. https://doi.org/10.1016/j.carbon.2011.11.010
45. Stankovich S, Dikin DA, Piner RD, Kohlhaas KA, Kleinhammes A, Jia Y, Wu Y, Nguyen SBT, Ruoff RS (2007) Synthesis of graphene-based nanosheets via chemical reduction of exfoliated graphite oxide. Carbon N Y 45:1558–1565. https://doi.org/10.1016/j.carbon.2007.02.034
46. Compton OC, Jain B, Dikin DA, Abouimrane A, Amine K, Nguyen ST (2011) Chemically active reduced graphene oxide with tunable C/O ratios. ACS Nano 5:4380–4391. https://doi.org/10.1021/nn1030725
47. Fernández-Merino MJ, Guardia L, Paredes JI, Villar-Rodil S, Solís-Fernández P, Martínez-Alonso A, Tascón JMD (2010) Vitamin C is an ideal substitute for hydrazine in the reduction of graphene oxide suspensions. J Phys Chem C 114:6426–6432. https://doi.org/10.1021/jp100603h
48. Moon IK, Lee J, Ruoff RS, Lee H (2010) Reduced graphene oxide by chemical graphitization. Nat Commun 1:1–6. https://doi.org/10.1038/ncomms1067
49. Shin HJ, Kim KK, Benayad A, Yoon SM, Park HK, Jung IS, Jin MH, Jeong HK, Kim JM, Choi JY, Lee YH (2009) Efficient reduction of graphite oxide by sodium borohydride and its effect on electrical conductance. Adv Funct Mater 19:1987–1992. https://doi.org/10.1002/adfm.200900167

50. Qiu T, Luo B, Liang M, Ning J, Wang B, Li X, Zhi L (2015) Hydrogen reduced graphene oxide/metal grid hybrid film: towards high performance transparent conductive electrode for flexible electrochromic devices. Carbon N Y 81:232–238. https://doi.org/10.1016/j.carbon.2014.09.054
51. Pham VH, Pham HD, Dang TT, Hur SH, Kim EJ, Kong BS, Kim S, Chung JS (2012) Chemical reduction of an aqueous suspension of graphene oxide by nascent hydrogen. J Mater Chem 22:10530–10536. https://doi.org/10.1039/c2jm30562c
52. Su CY, Xu Y, Zhang W, Zhao J, Tang X, Tsai CH, Li LJ (2009) Electrical and spectroscopic characterizations of ultra-large reduced graphene oxide monolayers. Chem Mater 21:5674–5680. https://doi.org/10.1021/cm902182y
53. Savchak M, Borodinov N, Burtovyy R, Anayee M, Hu K, Ma R, Grant A, Li H, Cutshall DB, Wen Y, Koley G, Harrell WR, Chumanov G, Tsukruk V, Luzinov I (2018) Highly conductive and transparent reduced graphene oxide nanoscale films via thermal conversion of polymer-encapsulated graphene oxide sheets. ACS Appl Mater Interfaces 10:3975–3985. https://doi.org/10.1021/acsami.7b16500
54. Wang X, Zhi L, Müllen K (2008) Transparent, conductive graphene electrodes for dye-sensitized solar cells. Nano Lett 8:323–327. https://doi.org/10.1021/nl072838r
55. Sun HB, Yang J, Zhou YZ, Zhao N, Li D (2014) Preparation of reduced graphene oxide films by dip coating technique and their electrical conductivity. Mater Technol 29:14–20. https://doi.org/10.1179/1753555712Y.0000000058
56. Wang X, Kholmanov I, Chou H, Ruoff RS (2015) Simultaneous electrochemical reduction and delamination of graphene oxide films. ACS Nano 9:8737–8743. https://doi.org/10.1021/acsnano.5b03814
57. Wang M, Duong LD, Oh JS, Mai NT, Kim S, Hong S, Hwang T, Lee Y, Do NJ (2014) Large-area, conductive and flexible reduced graphene oxide (RGO) membrane fabricated by electrophoretic deposition (EPD). ACS Appl Mater Interfaces 6:1747–1753. https://doi.org/10.1021/am404719u
58. Becerril HA, Mao J, Liu Z, Stoltenberg RM, Bao Z, Chen Y (2008) Evaluation of solution-processed reduced graphene oxide films as transparent conductors. ACS Nano 2:463–470. https://doi.org/10.1021/nn700375n
59. Kotsuki K, Tanaka H, Obata S, Stauss S, Terashima K, Saiki K (2014) The importance of spinning speed in fabrication of spin-coated organic thin film transistors: film morphology and field effect mobility. Appl Phys Lett 104. https://doi.org/10.1063/1.4883216
60. Obata S, Tanaka H, Saiki K (2013) Electrical and spectroscopic investigations on the reduction mechanism of graphene oxide. Carbon N Y 55:126–132. https://doi.org/10.1016/j.carbon.2012.12.018
61. Rivera Tito HA, Habermehl A, Müller C, Beck S, Romero Nieto C, Hernández Sosa G, Quintana Caceda ME (2017) Electrical and optical properties of reduced graphene oxide thin film deposited onto polyethylene terephthalate by spin coating technique. Appl Opt 56:7774. https://doi.org/10.1364/ao.56.007774
62. Graphene S, Wu J, Agrawal M, Becerril A, Bao Z, Liu Z, Chen ḰY, Peumans P (2010) Organic light-emitting diodes on solution-processed graphene transparent electrodes. ACS Nano 4:43–48
63. Zhao J, Pei S, Ren W, Gao L, Cheng HM (2010) Efficient preparation of large-area graphene oxide sheets for transparent conductive films. ACS Nano 4:5245–5252. https://doi.org/10.1021/nn1015506
64. Sun H, Chen D, Ye C, Li X, Dai D, Yuan Q, Chee KWA, Zhao P, Jiang N, Te LC (2018) Large-area self-assembled reduced graphene oxide/electrochemically exfoliated graphene hybrid

films for transparent electrothermal heaters. Appl Surf Sci 435:809–814. https://doi.org/10.1016/j.apsusc.2017.11.182
65. Hwang B, Park M, Kim T, Han SM (2016) Effect of RGO deposition on chemical and mechanical reliability of Ag nanowire flexible transparent electrode. RSC Adv 6:67389–67395. https://doi.org/10.1039/c6ra10338c
66. Pham VH, Cuong TV, Hur SH, Shin EW, Kim JS, Chung JS, Kim EJ (2010) Fast and simple fabrication of a large transparent chemically-converted graphene film by spray-coating. Carbon N Y 48:1945–1951. https://doi.org/10.1016/j.carbon.2010.01.062
67. Ricciardulli AG, Yang S, Wetzelaer GJAH, Feng X, Blom PWM (2018) Hybrid silver nanowire and graphene-based solution-processed transparent electrode for organic optoelectronics. Adv Funct Mater 28:1–6. https://doi.org/10.1002/adfm.201706010
68. Wen J, Jiang Y, Yang Y, Li S (2014) Conducting polymer and reduced graphene oxide Langmuir-Blodgett films: a hybrid nanostructure for high performance electrode applications. J Mater Sci Mater Electron 25:1063–1071. https://doi.org/10.1007/s10854-013-1687-z
69. Lin X, Jia J, Yousefi N, Shen X, Kim JK (2013) Excellent optoelectrical properties of graphene oxide thin films deposited on a flexible substrate by Langmuir-Blodgett assembly. J Mater Chem C Mater 1:6869–6877. https://doi.org/10.1039/c3tc31497a
70. De Silva KKH, Huang HH, Joshi RK, Yoshimura M (2017) Chemical reduction of graphene oxide using green reductants. Carbon N Y 119:190–199. https://doi.org/10.1016/j.carbon.2017.04.025
71. Pei S, Zhao J, Du J, Ren W, Cheng HM (2010) Direct reduction of graphene oxide films into highly conductive and flexible graphene films by hydrohalic acids. Carbon N Y 48:4466–4474. https://doi.org/10.1016/j.carbon.2010.08.006
72. Bae G, Cha J, Lee H, Park W, Park N (2012) Effects of defects and non-coordinating molecular overlayers on the work function of graphene and energy-level alignment with organic molecules. Carbon N Y 50:851–856. https://doi.org/10.1016/j.carbon.2011.09.044
73. Banhart F, Kotakoski J, Krasheninnikov AV (2011) Structural defects in graphene. ACS Nano 5:26–41. https://doi.org/10.1021/nn102598m
74. Ugeda MM, Brihuega I, Guinea F, Gómez-Rodríguez JM (2010) Missing atom as a source of carbon magnetism. Phys Rev Lett 104:1–4. https://doi.org/10.1103/PhysRevLett.104.096804
75. Stankovich S, Dikin DA, Piner RD, Kohlhaas KA, Kleinhammes A, Jia Y, Wu Y, Nguyen SBT, Ruoff RS (2007) Synthesis of graphene-based nanosheets via chemical reduction of exfoliated graphite oxide. Carbon N Y 45:1558–1565. https://doi.org/10.1016/j.carbon.2007.02.034
76. Seager CH, Pike GE (1974) Percolation and conductivity: a computer study. II. Phys Rev B 10:1435–1446. https://doi.org/10.1103/PhysRevB.10.1435
77. Vollmer A (2010) Introduction into X-ray and UV Photoelectron Spectroscopy (XPS/UPS). 61
78. Kumar PV, Bernardi M, Grossman JC (2013) The impact of functionalization on the stability, work function, and photoluminescence of reduced graphene oxide. ACS Nano 7:1638–1645. https://doi.org/10.1021/nn305507p
79. Long F, Yasaei P, Sanoj R, Yao W, Král P, Salehi-Khojin A, Shahbazian-Yassar R (2016) Characteristic work function variations of graphene line defects. ACS Appl Mater Interfaces 8:18360–18366. https://doi.org/10.1021/acsami.6b04853
80. Yan X, Cui X, Li B, Li LS (2010) Large, solution-processable graphene quantum dots as light absorbers for photovoltaics. Nano Lett 10:1869–1873. https://doi.org/10.1021/nl101060h
81. Chen W, Lv G, Hu W, Li D, Chen S, Dai Z (2018) Synthesis and applications of graphene quantum dots: a review. Nanotechnol Rev 7:157–185. https://doi.org/10.1515/ntrev-2017-0199
82. Ye R, Xiang C, Lin J, Peng Z, Huang K, Yan Z, Cook NP, Samuel ELG, Hwang CC, Ruan G, Ceriotti G, Raji ARO, Martí AA, Tour JM (2013) Coal as an abundant source of graphene quantum dots. Nat Commun 4. https://doi.org/10.1038/ncomms3943

83. Luo Z, Qi G, Chen K, Zou M, Yuwen L, Zhang X, Huang W, Wang L (2016) Microwave-assisted preparation of white fluorescent graphene quantum dots as a novel phosphor for enhanced white-light-emitting diodes. Adv Funct Mater 26:2739–2744. https://doi.org/10.1002/adfm.201505044
84. Gu S, Te HC, Chiang YM, Tzou DY, Chen YF, Gandomi YA (2018) Optimization of graphene quantum dots by chemical exfoliation from graphite powders and carbon nanotubes. Mater Chem Phys 215:104–111. https://doi.org/10.1016/j.matchemphys.2018.05.016
85. Lu Q, Wu C, Liu D, Wang H, Su W, Li H, Zhang Y, Yao S (2017) A facile and simple method for synthesis of graphene oxide quantum dots from black carbon. Green Chem 19:900–904. https://doi.org/10.1039/c6gc03092k
86. Peng J, Gao W, Gupta BK, Liu Z, Romero-Aburto R, Ge L, Song L, Alemany LB, Zhan X, Gao G, Vithayathil SA, Kaipparettu BA, Marti AA, Hayashi T, Zhu JJ, Ajayan PM (2012) Graphene quantum dots derived from carbon fibers. Nano Lett 12:844–849. https://doi.org/10.1021/nl2038979
87. Pan D, Zhang J, Li Z, Wu M (2010) Hydrothermal route for cutting graphene sheets into blue-luminescent graphene quantum dots. Adv Mater 22:734–738. https://doi.org/10.1002/adma.200902825
88. Shen J, Zhu Y, Chen C, Yang X, Li C (2011) Facile preparation and upconversion luminescence of graphene quantum dots. Chem Commun 47:2580–2582. https://doi.org/10.1039/c0cc04812g
89. Pousa Soares CP, de Lacerda Baptista R, Cesar DV (2018) Solvothermal reduction of graphite oxide using alcohols. Mater Res 21:1–7. https://doi.org/10.1590/1980-5373-MR-2017-0726
90. Zhu S, Song Y, Wang J, Wan H, Zhang Y, Ning Y, Yang B (2017) Photoluminescence mechanism in graphene quantum dots: quantum confinement effect and surface/edge state. Nano Today 13:10–14. https://doi.org/10.1016/j.nantod.2016.12.006
91. Tang L, Ji R, Cao X, Lin J, Jiang H, Li X, Teng KS, Luk CM, Zeng S, Hao J, Lau SP (2012) Deep ultraviolet photoluminescence of water-soluble self-passivated graphene quantum dots. ACS Nano 6:5102–5110. https://doi.org/10.1021/nn300760g
92. Kim R, Perebeinos V, Avouris P (2011) Relaxation of optically excited carriers in graphene. Phys Rev B Condens Matter Mater Phys 84:1–5. https://doi.org/10.1103/PhysRevB.84.075449
93. Zhu S, Shao J, Song Y, Zhao X, Du J, Wang L, Wang H, Zhang K, Zhang J, Yang B (2015) Investigating the surface state of graphene quantum dots. Nanoscale 7:7927–7933. https://doi.org/10.1039/c5nr01178g
94. Xu Q, Zhou Q, Hua Z, Xue Q, Zhang C, Wang X, Pan D, Xiao M (2013) Single-particle spectroscopic measurements of fluorescent graphene quantum dots. ACS Nano 7:10654–10661. https://doi.org/10.1021/nn4053342
95. Tetsuka H, Asahi R, Nagoya A, Okamoto K, Tajima I, Ohta R, Okamoto A (2012) Optically tunable amino-functionalized graphene quantum dots. Adv Mater 24:5333–5338. https://doi.org/10.1002/adma.201201930
96. Jin SH, Kim DH, Jun GH, Hong SH, Jeon S (2013) Tuning the photoluminescence of graphene quantum dots through the charge transfer effect of functional groups. ACS Nano 7:1239–1245. https://doi.org/10.1021/nn304675g
97. Lemme MC, Echtermeyer TJ, Baus M, Kurz H (2007) A graphene field-effect device. IEEE Electron Device Lett 28:282–284. https://doi.org/10.1109/LED.2007.891668
98. Park JK, Song SM, Mun JH, Cho BJ (2011) Graphene gate electrode for MOS structure-based electronic devices. Nano Lett 11:5383–5386. https://doi.org/10.1021/nl202983x
99. Yang N, Zhai J, Wang D, Chen Y, Jiang L (2010) Two-dimensional graphene bridges enhanced photoinduced charge transport in dye-sensitized solar cells. ACS Nano 4:887–894. https://doi.org/10.1021/nn901660v

100. Wu Z, Bai S, Xiang J, Yuan Z, Yang Y, Cui W, Gao X, Liu Z, Jin Y, Sun B (2014) Efficient planar heterojunction perovskite solar cells employing graphene oxide as hole conductor. Nanoscale 6:10505–10510. https://doi.org/10.1039/c4nr03181d
101. Lee KD, Park MJ, Kim DY, Kim SM, Kang B, Kim S, Kim H, Lee HS, Kang Y, Yoon SS, Hong BH, Kim D (2015) Graphene quantum dot layers with energy-down-shift effect on crystalline-silicon solar cells. ACS Appl Mater Interfaces 7:19043–19049. https://doi.org/10.1021/acsami.5b03672
102. Wang Q, Zhang X, Jin Z, Zhang J, Gao Z, Li Y, Liu SF (2017) Energy-down-shift CsPbCl3: Mn quantum dots for boosting the efficiency and stability of Perovskite solar cells. ACS Energy Lett 2:1479–1486. https://doi.org/10.1021/acsenergylett.7b00375
103. Zhang Q, Jie J, Diao S, Shao Z, Zhang Q, Wang L, Deng W, Hu W (2015) Solution-processed graphene quantum dot deep-UV. 1561–1570
104. Son D, Cho S, Nam J, Lee H, Kim M (2020) X-ray-based spectroscopic techniques for characterization of polymer nanocomposite materials at a molecular level. Polymers (Basel) 12. https://doi.org/10.3390/POLYM12051053
105. Kumar V, Kumar A, Lee DJ, Park SS (2021) Estimation of number of graphene layers using different methods: a focused review. Materials 14. https://doi.org/10.3390/ma14164590
106. Kastner J, Pichler T, Kuzmany H, Curran S, Blau W, Weldon DN, Delamesiere M, Draper S, Zandbergen H (1994) Resonance Raman and infrared spectroscopy of carbon nanotubes. Chem Phys Lett 221:53–58. https://doi.org/10.1016/0009-2614(94)87015-2
107. Guerrero-Contreras J, Caballero-Briones F (2015) Graphene oxide powders with different oxidation degree, prepared by synthesis variations of the Hummers method. Mater Chem Phys 153:209–220. https://doi.org/10.1016/j.matchemphys.2015.01.005
108. Bernard S, Beyssac O, Benzerara K, Findling N, Tzvetkov G, Brown GE (2010) XANES, Raman and XRD study of anthracene-based cokes and saccharose-based chars submitted to high-temperature pyrolysis. Carbon N Y 48:2506–2516. https://doi.org/10.1016/j.carbon.2010.03.024
109. Childres I, Jauregui LA, Park W, Caoa H, Chena YP (2013) Raman spectroscopy of graphene and related materials. In: New developments in photon and materials research, pp 403–418
110. Graf D, Molitor F, Ensslin K, Stampfer C, Jungen A, Hierold C, Wirtz L (2007) Spatially resolved raman spectroscopy of single- and few-layer graphene. Nano Lett 7:238–242. https://doi.org/10.1021/nl061702a
111. Silva DL, Campos JLE, Fernandes TFD, Rocha JN, Machado LRP, Soares EM, Miquita DR, Miranda H, Rabelo C, Vilela Neto OP, Jorio A, Cançado LG (2020) Raman spectroscopy analysis of number of layers in mass-produced graphene flakes. Carbon N Y 161:181–189. https://doi.org/10.1016/j.carbon.2020.01.050
112. Du D, Song H, Nie Y, Sun X, Chen L, Ouyang J (2015) Photoluminescence of graphene oxide in visible range arising from excimer formation. J Phys Chem C 119:20085–20090. https://doi.org/10.1021/acs.jpcc.5b04529
113. Gao W (2015) Graphene Oxide: Reduction Recipes, Spectroscopy, and Applications
114. Ickecan D, Zan R, Nezir S (2017) Eco-friendly synthesis and characterization of reduced graphene oxide. J Phys Conf Ser 902. https://doi.org/10.1088/1742-6596/902/1/012027
115. Krishnamoorthy K, Veerapandian M, Yun K, Kim SJ (2013) The chemical and structural analysis of graphene oxide with different degrees of oxidation. Carbon N Y 53:38–49. https://doi.org/10.1016/j.carbon.2012.10.013
116. Pei S, Cheng HM (2012) The reduction of graphene oxide. Carbon N Y 50:3210–3228. https://doi.org/10.1016/j.carbon.2011.11.010
117. University of Louisville (2013) Spin coating theory. https://louisville.edu/micronano/files/documents/standard-operating-procedures/SpinCoatingInfo.pdf/view 2–3

118. Shen B, Zhai W, Zheng W (2014) Ultrathin flexible graphene film: An excellent thermal conducting material with efficient EMI shielding. Adv Funct Mater 24:4542–4548. https://doi.org/10.1002/adfm.201400079
119. Chen C, Yang QH, Yang Y, Lv W, Wen Y, Hou PX, Wang M, Cheng HM (2009) Self-assembled free-standing graphite oxide membrane. Adv Mater 21:3007–3011. https://doi.org/10.1002/adma.200803726
120. Sun F, Ghosh H, Wang J, Tan Z, Sivoththaman S (2022) A two-step process for reduced graphene oxide films with work function tunability. IEEE Trans Nanotechnol 21:481–488. https://doi.org/10.1109/TNANO.2022.3193075
121. Yang H, Cao Y, He J, Zhang Y, Jin B, Sun JL, Wang Y, Zhao Z (2017) Highly conductive free-standing reduced graphene oxide thin films for fast photoelectric devices. Carbon N Y 115:561–570. https://doi.org/10.1016/j.carbon.2017.01.047
122. Behera S, Im J, Kim K (2020) Au-graphene oxide-anodic aluminum oxide nanostructured substrates for surface-enhanced Raman spectroscopy based molecular sensing. ACS Appl Nano Mater 3:914–922. https://doi.org/10.1021/acsanm.9b02575
123. Shen B, Zhai W, Zheng W (2014) Ultrathin flexible graphene film: an excellent thermal conducting material with efficient EMI shielding. Adv Funct Mater 24:4542–4548. https://doi.org/10.1002/adfm.201400079
124. Hontoria-Lucas C, López-Peinado AJ, López-González J de D, Rojas-Cervantes ML, Martín-Aranda RM (1995) Study of oxygen-containing groups in a series of graphite oxides: physical and chemical characterization. Carbon N Y 33:1585–1592. https://doi.org/10.1248/jhs1956.22.20
125. Ni Z, Wang Y, Yu T, Shen Z (2008) Raman spectroscopy and imaging of graphene. Nano Res 1:273–291. https://doi.org/10.1007/s12274-008-8036-1
126. Ferrari AC, Robertson J (2000) Interpretation of Raman spectra of disordered and amorphous carbon. A Phys Rev B 61:14095–14107. https://doi.org/10.1007/BF02543692
127. Moreira VR, Lebron YAR, da Silva MM, de Souza Santos LV, Jacob RS, de Vasconcelos CKB, Viana MM (2020) Graphene oxide in the remediation of norfloxacin from aqueous matrix: simultaneous adsorption and degradation process. Environ Sci Pollut Res 27:34513–34528. https://doi.org/10.1007/s11356-020-09656-6
128. Kovtun A, Jones D, Dell'Elce S, Treossi E, Liscio A, Palermo V (2019) Accurate chemical analysis of oxygenated graphene-based materials using X-ray photoelectron spectroscopy. Carbon N Y 143:268–275. https://doi.org/10.1016/j.carbon.2018.11.012
129. Kadakal Ç, Duman T, Ekinci R (2018) Thermal degradation kinetics of ascorbic acid, thiamine and riboflavin in rosehip (Rosa canina L) nectar. Food Sci Technol 38:667–673. https://doi.org/10.1590/1678-457x.11417
130. Schniepp HC, Li JL, McAllister MJ, Sai H, Herrera-Alonson M, Adamson DH, Prud'homme RK, Car R, Seville DA, Aksay IA (2006) Functionalized single graphene sheets derived from splitting graphite oxide. J Phys Chem B 110:8535–8539. https://doi.org/10.1021/jp060936f
131. Naghdi S, Sanchez-Arriaga G, Rhee KY (2019) Tuning the work function of graphene toward application as anode and cathode. J Alloys Compd 805:1117–1134. https://doi.org/10.1016/j.jallcom.2019.07.187
132. Sze SM, Kwok KNG (1981) Physics of semiconductor devices
133. Park Y, Choong V, Gao Y, Hsieh BR, Tang CW (1996) Work function of indium tin oxide transparent conductor measured by photoelectron spectroscopy. Appl Phys Lett 68:2699–2701. https://doi.org/10.1063/1.116313
134. Li J, Qi X, Hao G, Ren L, Zhong J (2015) In-situ investigation of graphene oxide under UV irradiation: evolution of work function. AIP Adv 5. https://doi.org/10.1063/1.4923238

135. Park Y, Choong V, Gao Y, Hsieh BR, Tang CW (1996) Work function of indium tin oxide transparent conductor measured by photoelectron spectroscopy. Appl Phys Lett 68:2699–2701. https://doi.org/10.1063/1.116313
136. Abid SP, Islam SS, Mishra P, Ahmad S (2018) Reduced graphene oxide (rGO) based wideband optical sensor and the role of temperature, defect states and quantum efficiency. Sci Rep 8:1–13. https://doi.org/10.1038/s41598-018-21686-2
137. Eda G, Lin YY, Mattevi C, Yamaguchi H, Chen HA, Chen IS, Chen CW, Chhowalla M (2010) Blue photoluminescence from chemically derived graphene oxide. Adv Mater 22:505–509. https://doi.org/10.1002/adma.200901996
138. Chien CT, Li SS, Lai WJ, Yeh YC, Chen HA, Chen IS, Chen LC, Chen KH, Nemoto T, Isoda S, Chen M, Fujita T, Eda G, Yamaguchi H, Chhowalla M, Chen CW (2012) Tunable photoluminescence from graphene oxide. Angew Chem Int Edn 51:6662–6666. https://doi.org/10.1002/anie.201200474
139. Waremra RS, Betaubun P (2018) Analysis of electrical properties using the four point probe method. E3S Web Conf 73:1–4. https://doi.org/10.1051/e3sconf/20187313019
140. Misra A, Kalita H, Kottantharayil A (2014) Work function modulation and thermal stability of reduced graphene oxide gate electrodes in MOS devices. ACS Appl Mater Interfaces 6:786–794. https://doi.org/10.1021/am404649a
141. Xueshen W, Jinjin L, Qing Z, Yuan Z, Mengke Z (2013) Thermal annealing of exfoliated graphene. J Nanomater 2013. https://doi.org/10.1155/2013/101765
142. Ganguly A, Sharma S, Papakonstantinou P, Hamilton J (2011) Probing the thermal deoxygenation of graphene oxide using high-resolution in situ X-ray-based spectroscopies. J Phys Chem C 115:17009–17019. https://doi.org/10.1021/jp203741y
143. Becker W (2010) The bh TCSPC handbook, pp 1–566
144. Sarkar S, Gandla D, Venkatesh Y, Bangal PR, Ghosh S, Yang Y, Misra S (2016) Graphene quantum dots from graphite by liquid exfoliation showing excitation-independent emission, fluorescence upconversion and delayed fluorescence. Phys Chem Chem Phys 18:21278–21287. https://doi.org/10.1039/c6cp01528j
145. Dong Y, Zhang S, Shi L, Chen Y, Ma J, Guo S, Chen X, Song H (2018) The photoluminescence of step-wise reduced graphene oxide quantum dots. Mater Chem Phys 203:125–132. https://doi.org/10.1016/j.matchemphys.2017.09.064
146. Zhu X, Xiao X, Zuo X, Liang Y, Nan J (2014) Hydrothermal preparation of photoluminescent graphene quantum dots characterized excitation-independent emission and its application as a bioimaging reagent. 801–809. https://doi.org/10.1002/ppsc.201300375
147. Bruna M, Ott AK, Ijäs M, Yoon D, Sassi U, Ferrari AC (2014) Doping dependence of the Raman spectrum of defected graphene. ACS Nano 8:7432–7441. https://doi.org/10.1021/nn502676g
148. Li H, He X, Kang Z, Huang H, Liu Y, Liu J, Lian S, Tsang CHA, Yang X, Lee ST (2010) Water-soluble fluorescent carbon quantum dots and photocatalyst design. Angew Chem Int Edn 49:4430–4434. https://doi.org/10.1002/anie.200906154
149. Wang L, Zhu SJ, Wang HY, Qu SN, Zhang YL, Zhang JH, Chen QD, Xu HL, Han W, Yang B, Sun HB (2014) Common origin of green luminescence in carbon nanodots and graphene quantum dots. ACS Nano 8:2541–2547. https://doi.org/10.1021/nn500368m
150. Savintsev AP, Gavasheli YO, Kalazhokov ZK, Kalazhokov KK (2016) X-ray photoelectron spectroscopy studies of the sodium chloride surface after laser exposure. J Phys Conf Ser 774. https://doi.org/10.1088/1742-6596/774/1/012118
151. Li Y, Liu X, Wang J, Liu H, Li S, Hou Y, Wan W, Xue W, Ma N, Zhang JZ (2016) Chemical nature of redox-controlled photoluminescence of graphene quantum dots by post-synthesis treatment. J Phys Chem C 120:26004–26011. https://doi.org/10.1021/acs.jpcc.6b07935

152. Kagan CR, Lifshitz E, Sargent EH, Talapin DV (2016) Building devices from colloidal quantum dots. Science 1979:353. https://doi.org/10.1126/science.aac5523
153. Boukhvalov DW, Katsnelson MI (2009) Chemical functionalization of graphene with defects. J Phys Condens Matter 21. https://doi.org/10.1088/0953-8984/21/34/344205
154. Cai J, Ruffieux P, Jaafar R, Bieri M, Braun T, Blankenburg S, Muoth M, Seitsonen AP, Saleh M, Feng X, Müllen K, Fasel R (2010) Atomically precise bottom-up fabrication of graphene nanoribbons. Nature 466:470–473. https://doi.org/10.1038/NATURE09211
155. Zhang BY, Liu T, Meng B, Li X, Liang G, Hu X, Wang QJ (2013) Broadband high photoresponse from pure monolayer graphene photodetector. Nat Commun 4:1–11. https://doi.org/10.1038/ncomms2830
156. Yuan M, Liu M, Sargent EH (2016) Colloidal quantum dot solids for solution-processed solar cells. Nat Energy 1. https://doi.org/10.1038/nenergy.2016.16
157. Zhang Y, Zherebetskyy D, Bronstein ND, Barja S, Lichtenstein L, Schuppisser D, Wang LW, Alivisatos AP, Salmeron M (2015) Charge percolation pathways guided by defects in quantum dot solids. Nano Lett 15:3249–3253. https://doi.org/10.1021/acs.nanolett.5b00454
158. Nagpal P, Klimov VI (2011) Role of mid-gap states in charge transport and photoconductivity in semiconductor nanocrystal films. Nat Commun 2. https://doi.org/10.1038/ncomms1492
159. Wang ZM, Salamo G, Bellucci S (2021) Quantum Dot Photodetectors
160. Konstantatos G, Sargent EH (2011) Colloidal quantum dot photodetectors. Infrared Phys Technol 54:278–282. https://doi.org/10.1016/j.infrared.2010.12.029
161. Schlaf R, Murata H, Kafafi ZH (2001) Work function measurements on indium tin oxide films. J Electron Spectros Relat Phenomena 120:149–154. https://doi.org/10.1016/S0368-2048(01)00310-3
162. Al-Kuhaili MF (2020) Electrical conductivity enhancement of indium tin oxide (ITO) thin films reactively sputtered in a hydrogen plasma. J Mater Sci: Mater Electron 31:2729–2740. https://doi.org/10.1007/s10854-019-02813-9
163. Lim HC, Min SH, Lee E, Jang J, Kim SH, Hong JI (2015) Self-assembled Poly(3,4-ethylene dioxythiophene):Poly(styrenesulfonate)/graphene quantum dot organogels for efficient charge transport in photovoltaic devices. ACS Appl Mater Interfaces 7:11069–11073. https://doi.org/10.1021/acsami.5b02434
164. Bhujel R, Rai S, Deka U, Sarkar G, Biswas J, Swain BP (2021) Bandgap engineering of PEDOT:PSS/rGO a hole transport layer for SiNWs hybrid solar cells. Bull Mater Sci 44. https://doi.org/10.1007/s12034-021-02376-8
165. Valtakari D, Liu J, Kumar V, Xu C, Toivakka M, Saarinen JJ (2015) Conductivity of PEDOT:PSS on spin-coated and drop cast nanofibrillar cellulose thin films. Nanoscale Res Lett 10:1–10. https://doi.org/10.1186/s11671-015-1093-y
166. Du FP, Cao NN, Zhang YF, Fu P, Wu YG, Lin ZD, Shi R, Amini A, Cheng C (2018) PEDOT:PSS/graphene quantum dots films with enhanced thermoelectric properties via strong interfacial interaction and phase separation. Sci Rep 8:1–12. https://doi.org/10.1038/s41598-018-24632-4
167. Kim HJ, Lee CK, Seo JG, Hong SJ, Song G, Lee J, Ahn C, Lee DJ, Song SH (2020) Highly luminescent polyethylene glycol-passivated graphene quantum dots for light emitting diodes. RSC Adv 10:27418–27423. https://doi.org/10.1039/d0ra02257h
168. Wang S, Cole IS, Zhao D, Li Q (2016) The dual roles of functional groups in the photoluminescence of graphene quantum dots. Nanoscale 8:7449–7458. https://doi.org/10.1039/c5nr07042b
169. Lou Y, Ji J, Qin A, Liao L, Li Z, Chen S, Zhang K, Ou J (2020) Cane molasses graphene quantum dots passivated by PEG functionalization for detection of metal ions. ACS Omega 5:6763–6772. https://doi.org/10.1021/acsomega.0c00098

170. Park M, Jeong Y, Kim HS, Lee W, Nam SH, Lee S, Yoon H, Kim J, Yoo S, Jeon S (2021) Quenching-resistant solid-state photoluminescence of graphene quantum dots: reduction of $\pi-\pi$ stacking by surface functionalization with POSS, PEG, and HDA. Adv Funct Mater 31:1–10. https://doi.org/10.1002/adfm.202102741
171. Yin X, Zhang C, Guo Y, Yang Y, Xing Y, Que W (2021) PbS QD-based photodetectors: future-oriented near-infrared detection technology. J Mater Chem C Mater 9:417–438. https://doi.org/10.1039/d0tc04612d
172. Zhu S, Song Y, Wang J, Wan H, Zhang Y, Ning Y, Yang B (2017) Photoluminescence mechanism in graphene quantum dots: quantum confinement effect and surface/edge state. Nano Today 13:10–14. https://doi.org/10.1016/j.nantod.2016.12.006
173. Lundgren M, Allan NL, Cosgrove T, George N (2002) Wetting of water and water/ethanol droplets on a non-polar surface: a molecular dynamics study. Langmuir 18:10462–10466. https://doi.org/10.1021/la026191w
174. Zheng Y (2021) Scalable synthesis of less-toxic colloidal copper indium sulfide quantum dots and application to optoelectronic device architectures by
175. Lin L, Peng H, Liu Z. Synthesis challenges for graphene industry

GPSR Compliance

The European Union's (EU) General Product Safety Regulation (GPSR) is a set of rules that requires consumer products to be safe and our obligations to ensure this.

If you have any concerns about our products, you can contact us on ProductSafety@springernature.com

In case Publisher is established outside the EU, the EU authorized representative is:

Springer Nature Customer Service Center GmbH
Europaplatz 3
69115 Heidelberg, Germany

Batch number: 10732638

Printed by Printforce, the Netherlands